# Welcome to AQA GCSE Science!

## Learning objectives

Each topic begins with key questions that you should be able to answer by the end of the lesson.

 **Examiner's tip**

AQA Examiner's tip are hints from the examiners who will mark your exams, giving you important advice on things to remember and what to watch out for.

 **Did you know ...?**

There are lots of interesting and often strange things about Science. This feature tells you about many of them.

## ∞ links

*Links will tell you where you can find more information about what you're learning.*

## Activity

Activity is linked to a main lesson and could be a discussion or task in pairs, groups or by yourself.

 **Maths skills**

This feature highlights the maths skills that you will need for your Science exams with short, visual explanations.

---

This book has been written for you by the people who will be marking your exams, very experienced teachers and subject experts. It covers everything you need to know for your exams and is packed full of features to help you achieve the very best that you can.

Questions in yellow boxes check that you understand what you're learning as you go along. The answers are all within the text so if you don't know the answer, you can go back and re-read the relevant section

**Figure 1** Many diagrams are as important for you to learn as the text, so make sure you revise them carefully.

Key words are highlighted in the text and are shown **like this**. You can look them up in the glossary at the back of the book if you're not sure what they mean.

**k** Where you see this icon, there are supporting electronic resources in our kerboodle! online service.

## Practical

This feature helps you become familiar with key practicals. It may be a simple introduction, reminder or the basis for a practical in the classroom.

Anything in the Higher Tier boxes must be learned by those sitting the Higher Tier exam. If you'll be sitting the Foundation Tier, these boxes can be missed out.

The same is true for any other places which are marked Higher or **[H]**.

**Higher**

## Summary questions

Summary questions give you the chance to test whether you have learned and understood everything in the topic. If you get any wrong, go back and have another look.

---

And at the end of each chapter you will find …

## Summary questions

These questions will test you on what you have learned throughout the whole chapter, helping you to work out what you have understood and where you need to go back and revise.

## AQA Examination-style questions

These questions are examples of the types of questions you will answer in your actual GCSE, so you can get lots of practice during your course.

## Key points

at the end of the topic are the important points that you must remember. They can be used to help revision and summarise your knowledge.

# My wider world

In Unit 1, Theme 1 you will work in the following contexts, covered in Chapters 1 and 2:

## Our changing planet and universe

### How do we learn about other parts of the universe?

Light from distant stars has taken millions of years to reach us. So, when we see a star, we are actually seeing it as it was a long time ago. Telescopes based on Earth or in space help scientists discover what the stars and universe were like in the past. They detect different forms of electromagnetic radiation so we can study distant stars, black holes and other objects even further back in time.

### Why do we think the universe is changing?

Scientists study light from distant stars. They can say what elements are present in stars billions of miles away and how fast stars and galaxies are moving. Studying the light from distant galaxies helps astronomers explain changes in the universe that are happening now, as well as modelling how the universe began.

### What do we know about the Earth's place in the universe?

The Sun seems to move through the sky as the day goes on. Over many centuries, astronomers have studied how the Earth and its neighbouring planets orbit our star, the Sun.

## Our changing planet

### How has the Earth changed since it formed?

The landscape around us does not seem to change. If we look a bit closer we can tell that things are not as they seem. Rocks have fossils in them. These rocks have been crushed and bent into odd shapes. The Earth has cooled and continues to change since it formed billions of years ago. Its structure is layered and cracked near the surface. Tectonic plates move constantly so that continents and oceans are now dramatically different from when they first formed.

### What causes the changes on Earth?

Convection currents in the molten rock under the Earth's surface force the tectonic plates to move. This is the cause of earthquakes and volcanoes that can devastate large areas.

## Has our atmosphere always been the same?

We take the air for granted but did you know that we are only here today because of microorganisms? When the Earth formed, gases were released during the many violent eruptions from volcanoes that were active at that time. Water vapour condensed to form oceans. The remaining gases somehow allowed single-celled organisms to grow. Some of these produced the oxygen now in the atmosphere.

### Why can the Earth support life?

Our atmosphere surrounds the Earth and the layer of gases helps to sustain life. It plays a vital part in keeping temperatures on Earth stable. It allows some energy from the Sun through. It also absorbs some radiation going out from the Earth.

## Materials our planet provides

### Raw materials

Mining and quarrying extract millions of tonnes of material from the Earth every year. Some of these materials are elements, some are compounds and some are mixtures. Mixtures like crude oil or rock salt are quite easy to separate. This is because the substances in them are not chemically joined. Compounds, found in mineral ores, are harder to separate, because the elements they are made of are chemically joined. Some materials such as gold or sulfur exist as elements in the Earth's crust. Not all materials we use come from the Earth's crust. For example, gases such as nitrogen are extracted from the atmosphere.

## Using materials from our planet to make products

### Making products

Most of the time, raw materials need to be chemically changed before they can be used. This means that knowledge of chemical reactions is vital when making new products. To make it easier to communicate about chemical reactions, they can be written as equations. Chemists can work out exactly how much of each raw material they need from balanced chemical equations. If they get it wrong, money is wasted and costs go up for us.

*How has space exploration helped us? New materials used for space missions have everyday uses too. For example, prosthetics (artificial limbs) are lighter and tougher. Our homes can be insulated using much thinner but more effective materials. Fibreglass roofing fabric is being used as a stunning new building material and improved thermal protection on racing cars keeps the drivers safer.*

*Volcanoes and earthquakes can have huge effects on our everyday life. In 2010 the ash from a volcano in Iceland halted flights all over the world.*

*Weather forecasters, use information from satellites, as well as taking readings from weather stations to predict the weather. We can use satellites to monitor the weather locally and different satellites to track weather systems such as cyclones. When you hear the weather forecast in the morning it will be based on information from satellites.*

*As well as the constant need for raw materials, getting new materials from the Earth's crust keeps many people employed. Without quarries there wouldn't be much employment in some communities. There are many skilled jobs in a quarry, such as blasting, monitoring the environment, running machinery and transporting the raw materials.*

## 1.1

# Observing our solar system

### Learning objectives

- How can we observe the solar system and the galaxies in the universe?

- What types of radiation can be detected by telescopes?

## How observing the universe has developed

For centuries, people could only look at **stars** using their eyes. Astronomers drew star charts, attempting to show the movements of the stars and their patterns. They could not explain exactly what they saw as the stars seemed too small to see any details.

When Galileo started using the **telescope** in the 1600s, he could see moons circling around nearby planets. He could also see Saturn's rings and the stars in our galaxy, the Milky Way. Since then, other scientists have improved the design of telescopes. Now we can observe stars that were formed early in the history of the universe. Their light has taken billions of years to reach us.

The problem with looking at a distant object is that very little light from it reaches us. Simply making an object's image larger can make it too dim to see. Telescopes use lenses and mirrors. They collect more light than our eyes can. They focus this light to make a brighter image. They also make the image bigger.

Our telescopes now are more powerful than ever before. For example, the Hubble Space Telescope shows galaxies and stars as they were 13 billion years ago.

a How do telescopes help us to see distant objects?

Telescopes can be based on the ground or in space. Telescopes on the ground are easier to use. They can be updated, maintained and visited easily. However, our atmosphere can spoil the images. For example, clouds can block our view of the sky.

Telescopes based in space take much clearer images because there is no light pollution or atmosphere present. However, they are expensive to run and very hard to mend or visit. You will see how the choice of telescope depends on different factors.

b Write down one advantage and one disadvantage of ground- and space-based telescopes.

**Figure 1** Telescopes are needed to study distant stars in more detail

### Different types of electromagnetic radiation

We can feel the Sun's warmth and it tans our skin. The Sun emits (gives out) light. It also, like other stars, emits infrared radiation, which warms us, and ultraviolet radiation, which tans our skin.

Modern telescopes detect light as well as other forms of radiation from stars. The images they make help astronomers find out even more information about distant objects.

Different stars emit different types of electromagnetic radiation. The type of radiation emitted by stars depends on their temperature. This affects the choice of telescope as some forms of radiation are absorbed by our atmosphere.

**Radio waves** are caused by the coolest sources – such as some stars, and planets like the Earth. Since radio waves can pass through our atmosphere, large **ground-based telescopes** detect these signals.

### AQA Examiner's tip

A satellite telescope is only about 60 miles up in space. Our nearest star is $3.97 \times 10^{13}$ km away. So, even though a satellite telescope is nearer the stars than one on the Earth's surface, it is not the shorter distance that makes the difference to the quality of the image. Satellite telescopes are put into space so that they are above clouds, light pollution and atmospheric pollution.

# AQA Science

## Science B: Science in Context

New GCSE

**James Hayward**

**Jo Locke**

**Nicky Thomas**

Louise Burt

Andrea Johnson

Series Editor
**Lawrie Ryan**

Nelson Thornes

# GCSE Science B: Science in Context

# Contents

507.6

Newly formed stars and cool dust clouds emit low-energy **microwaves**. The dust clouds contain tiny solid particles, and are seen as dark patches between stars. Ground-based telescopes also detect a background of microwaves coming from all directions throughout the universe. These microwaves were created at the Big Bang, the very start of the universe itself.

Some planets, stars such as the Sun and dust clouds emit **infrared radiation** – very hot stars emit **ultraviolet radiation**. Very little infrared or ultraviolet radiation from distant stars and galaxies travels through our atmosphere. This means **space-based telescopes** are needed to see objects giving out infrared and ultraviolet radiation. The Hubble Space Telescope is mainly designed to look at visible light, but can also detect some infrared and ultraviolet radiation.

X-rays and gamma rays are only detected by space-based telescopes as our atmosphere absorbs this radiation. **X-rays** are given out during supernovas – the huge explosions caused by the collapse of very massive stars as their lifecycle ends. X-rays are also given out by the uppermost level of the Sun's atmosphere. **Gamma rays** are the most energetic rays. We can see the remains of supernovas, neutron stars and black holes, by using telescopes to see the gamma rays they emit.

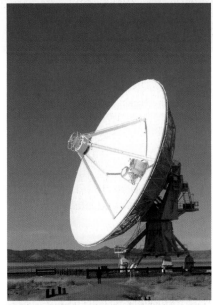

**Figure 2** Radio telescopes can be ground based as radio waves are not affected by our atmosphere

**Figure 3** Galaxies like this one emit gamma rays as they produce stars at their core

### Did you know ... ?

Anyone can apply to use the Hubble Space Telescope to view the universe. Hubble sees light from the most distant stars. Their light has taken so long to reach us that we see these stars as they appeared billions of years ago.

**c** Explain what affects the different types of electromagnetic radiation given out by different objects in space.

### links

*For more information on the electromagnetic spectrum see 9.6 Electromagnetic waves.*

## Gathering data

**Space probes** are spacecraft carrying scientific instruments. They are designed to gather data and send it back to Earth. The probes travel through the solar system and may land on other planets. They are not designed to return to Earth and so they can explore much further than other telescopes.

### Key points

- Telescopes are used to see light and other radiation coming from distant objects.
- Telescopes can be based on the ground or in space.
- Different telescopes see different types of electromagnetic radiation, allowing us to see different objects in space.

### Summary questions

1 Describe the main differences between space- and ground-based telescopes.

2 Explain why we use space probes and space-based telescopes, as well as ground-based telescopes.

3 Make a list of objects in space so that each emits a different type of electromagnetic radiation.

<table>
<tr><td>

**1.2**

</td><td>

# How the universe began

</td></tr>
</table>

## Learning objectives

- What happens to the observed wavelength and frequency of a moving wave source?

- What do we mean by 'the red shift' in light observed from most distant galaxies?

- What evidence do we have that the universe is expanding and how does this support the Big Bang theory?

## Wavelengths and wave source

Last century, Edwin Hubble used telescopes to measure the light from distant stars and galaxies. Light from stars is a blend of different colours. Different colours of light have a different blend of colours present. **Spectroscopy** shows these different colours of light as lines on a coloured spectrum. The blend of colours of light changes depending on which elements are present in the star.

**a** How can you tell which elements are present in a star?

Hubble used the pattern of lines to identify the elements in galaxies. The pattern of lines matched the elements Hubble expected to see. However, all lines were moved along the spectrum towards the redder light. More distant galaxies had a larger **red-shift** than closer ones.

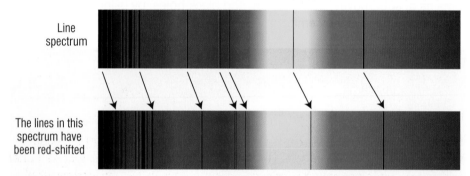

Line spectrum

The lines in this spectrum have been red-shifted

**Figure 1** A line spectrum contrasted with lines that have been red-shifted

To explain this, imagine that a train travels past, moving very fast. As it comes closer, the pitch of the noise from its engines rises. As it moves away again, the pitch falls. This is the **Doppler effect**.

The Doppler effect happens because a wave source moving towards us will squash the waves. When the wave source moves away, it stretches the waves. If the wavelength is stretched, then the frequency is lower. It happens with light waves as well as sound waves.

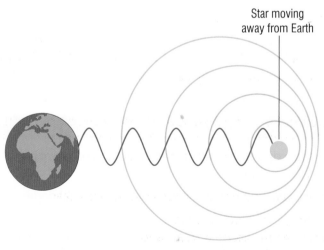

Star moving away from Earth

**Figure 2** The Doppler effect means that waves are stretched apart if a star moves away from the Earth

## ⬭⬭ links

*For more information on the frequency and wavelength of waves see 9.5 How fast do waves travel?*

**b** How can we tell that the stars are moving?

Red light has a longer wavelength than blue light. When a star or galaxy moves away from us, its light appears redder than expected because its wavelength is stretched. Hubble's results were evidence that galaxies move away from us. The galaxies further away have a bigger red shift because they are moving away faster than galaxies closer to us.

**c** How could Hubble tell that galaxies further away are moving faster than closer galaxies?

## The universe is expanding

One puzzle was that galaxies in all directions are moving away from the Earth. Does this mean the Earth is at the centre of the universe? Scientists believe that this is not the case. It is like blowing up a balloon with spots marked all over it. As it gets bigger, all the spots move away from each other. Similarly, if the whole universe is expanding, then galaxies will move away from each other.

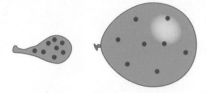

**Figure 3** As the balloon gets bigger, the spots spread apart

If the universe is getting bigger, then it must have been smaller in the past. In fact, many scientists believe that the universe was once so small that all matter and energy were squashed in one tiny place. This is called the 'cosmic egg'. At some point there was a huge explosion that scientists called the **Big Bang**. As the matter was flung outwards, the universe started expanding. It has continued to expand for 14 billion years and is still expanding today.

There is more evidence that the universe began as an explosion. Scientists using a radio telescope in the 1960s detected a weak signal coming from all directions, which they called **cosmic background radiation**. This was caused by microwave radiation left over from the Big Bang. It is often called the 'echo of the Big Bang' as it is what remains of the radiation created at that time.

**d** What evidence is there that the Big Bang took place?

### Did you know ...?

The researchers who discovered the cosmic background radiation were investigating something completely different. They thought the noise was caused by pigeon droppings in the receiver of the telescope.

## Summary questions

1 Copy and complete using the words below:

*Big Bang   Doppler effect   red-shift*

   **a** We believe the universe began as a big explosion called the ..............
   **b** The .............. is evidence that the universe is still expanding.
   **c** The .............. means that the wavelength of light stretches when a galaxy moves away from us.

2 Explain how the red-shift and the Doppler effect are linked.

3 Write down **one** piece of evidence that:
   **a** the universe began billions of years ago
   **b** the universe is still expanding.

## Key points

- The universe began as a huge explosion called the Big Bang.
- The red-shift is evidence that the universe is still expanding.
- The cosmic background radiation is evidence that the Big Bang took place.
- The wavelength and frequency of a moving light source change.

# 1.3 Our place in the universe

- What makes up our solar system?
- Which galaxy does our solar system belong to?
- Why is the position of the Earth in the solar system important?

Several thousand years ago, Egyptian astronomers noticed fixed patterns of **stars** like a belt across the sky. They called this the Zodiac belt. They noticed stars wandering in the night sky that were really **planets** orbiting the Sun.

An orbit is the path of an object, such as a planet, around a larger object, such as a star. Stars are balls of extremely hot, glowing gas. Planets orbit a central star and are visible because they reflect the star's light. Without telescopes, we can't see the difference between stars and planets.

**Figure 1** Without a telescope, we can't tell stars from planets, or how far different stars are from us. Three of the objects in the sky shown here are planets, the others are stars.

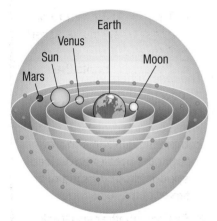

**Figure 2** Aristotle's early model of the solar system, with the Earth at the centre and the stars suspended on crystal spheres

## The Earth's position in the solar system

About 2500 years ago, Philolaus believed the Earth, Moon and stars orbited a central fire. However, Aristotle believed the Sun, Moon and planets orbited the Earth with stars suspended on crystal spheres. Ptolemy explained Aristotle's ideas about 300 years later.

About 500 years ago, Copernicus published a model of the **solar system** showing the Sun at the centre orbited by the Earth and other planets. The **Moon** orbited the Earth. Kepler realised the planets' **orbits** were actually ellipses (squashed circles). We use this model today.

About 400 years ago, Galileo used telescopes to observe **moons** orbiting Jupiter and rings around Saturn. Nowadays, information from telescopes confirm Galileo's model. More planets have been discovered since then. The furthest of these, Pluto, is now classified as a dwarf planet. Astronomers are still finding more dwarf planets.

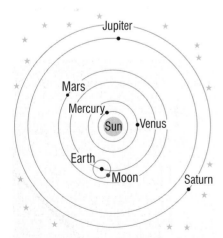

**Figure 3** Copernicus' model of the solar system, with the Sun positioned at the centre

   **a** What is the difference between a star and a planet?

a

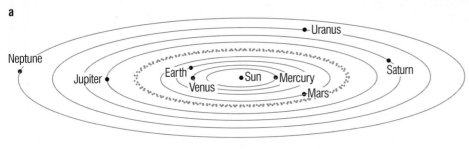

b

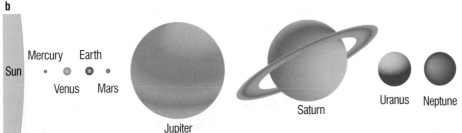

**Figure 4 a** The modern model of the solar system, where eight planets orbit the sun on ellipses
**b** The relative sizes of the Sun and planets

**b** Write down the main difference between the early Greek model of the solar system and Copernicus' model.

## The evolution of the universe

Our position on Earth is very special. For life, we need liquid water. If a planet orbits the Sun too closely, any water evaporates away. If it orbits too far away, all water would be frozen solid. We are the right distance from the Sun for water to be a liquid, sustaining life.

Earth isn't the only place with liquid water. Space probes have found water underneath an ice crust on one of Jupiter's moons. However, signs of life such as bacteria or other single-celled organisms were not found.

**c** What is special about the Earth's position in the solar system?

There are hundreds of billions of stars just in our **galaxy**, the Milky Way. In the universe, there could be hundreds of billions of galaxies. It is likely that there are billions of planets in the universe like Earth. The SETI institute was set up to look for extraterrestrial life.

Looking for extraterrestrial life is not easy. It takes years for space probes to travel to our neighbouring planets. A radio signal takes over four years to reach our nearest star. A reply would take nearly nine years to arrive. Scientists must decide where to send signals and what to send.

**d** Why do we think there could be life in other places in the universe?

## Summary questions

1 Copy and complete using the words below:

*Milky Way   Moon   Sun*

   **a** Our galaxy is called the ................

   **b** The star at the centre of our solar system is called the .............

   **c** The ............ orbits the Earth.

   **d** Our nearest star is called the .............

2 Create a timeline showing how our ideas of the solar system have changed over time.

3 Explain whether you think we should look for life somewhere else in the universe? What are the potential benefits and drawbacks?

## Key points

- Our models of the solar system have changed over centuries.

- The Earth orbits the Sun, which is at the centre of our solar system, which is in the Milky Way galaxy.

- It is likely there are other places in the galaxy that can support life.

## 1.4 The Earth's structure

### Learning objectives

- How has the Earth changed over time?
- What is the Earth's structure?

**Figure 1** Dr James Hutton was the first to suggest the Earth changes very slowly

### The age of the Earth

Three hundred years ago, scientists believed the Earth was only 6000 years old. There was good reason for this, of course. There were few historical records dating back this far, and an age of 6000 years agreed with religious texts. In the 1780s, a Scottish doctor called James Hutton challenged this belief.

Hutton spent a lot of time looking at rocks and identifying what they were made of. He started to see evidence that natural recycling was taking place, and wrote:

> A vast proportion of present rocks are composed of materials produced by the destruction of animal, vegetable and mineral substances of more ancient origin.

**a** About 300 years ago, how old did most scientists think the Earth was?

### The Earth's composition

**Geologists** study the Earth, measuring its composition and structure. They try to explain the processes that change the Earth. This is very valuable information to companies that mine the Earth for materials such as **metal ores** or **fossil fuels**. Knowing the types of rock formations that might contain precious metals or crude oil is worth a lot of money. A career in geology can involve a lot of travel around the world and work outdoors.

By measuring levels of **radioactivity** in rocks, we now know the Earth to be around 4.5 billion years old. Some major changes have happened during that time. The Earth looks very different from the way it did 4.5 billion years ago.

When the Earth had just formed, it was a ball of **molten** rock in constant motion. The dense metals iron and nickel started to sink into the centre of this ball. They are still hot billions of years later. Lighter elements around the outside of the Earth then cooled down, forming a solid crust. Its surface was very unstable and covered in **volcanoes**.

**Figure 2** Geologists at work taking a core sample – this helps build up a picture of the structure of the Earth's crust

**Figure 3** As the Earth cooled, the materials in it started to form layers

**b** Why does the core of the Earth contain iron and nickel?

We now know a lot about the structure of the Earth. Firstly, we know its **core** is a mixture of iron and nickel, which is solid in the middle, but liquid in the outer part of the core. Surrounding the core is a layer called the **mantle**. The mantle has many of the properties of a solid, but it can flow slowly in parts. This movement in the mantle can have dramatic and devastating consequences for us.

On top of the mantle is a very thin **crust** of solid rock, upon which we live. This solid crust is surrounded by a layer of gases we call our atmosphere.

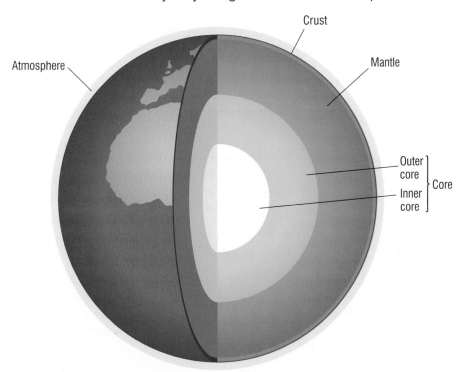

**Figure 4** The structure of the Earth

**??** **Did you know …?**

The Earth's crust is very thin. If the world was the size of a football, the crust would be as thick as a postage stamp on its surface.

## Summary questions

1 Copy and complete using the words below:

*core   crust   atmosphere   rock   mantle   flows*

The surface of the Earth is called the ............. Underneath this is a layer of ............. which we call the ............. This layer ............. very slowly over time. The centre of the Earth is called the ............. The Earth is surrounded by its .............

2 What skills and qualities do you think a geologist needs?

## Key points

- The .......... ball .......... slow .......... a cru..........

- Toda.......... a me.......... by a.......... gase..........

12

## 1.5

# Changes in the Earth's surface

### Learning objectives

- What are tectonic plates?
- What causes tectonic plates to move?
- What effects do movement of the Earth's tectonic plates cause?
- Can scientists predict earthquakes and volcanic eruptions?

Have you ever heated soup in a saucepan? If you leave it for a few minutes, you will see a skin starts to form on the surface. If you keep watching, you may see cracks or wrinkles start to appear, and very slight movements in the skin. The interior structure of the Earth behaves in a very similar way to your saucepan of soup.

Just like the soup in your saucepan, the Earth's mantle is being heated strongly from below. The heat is coming mainly from nuclear reactions in the core, as unstable elements go through radioactive decay. The heat causes **convection currents** in the mantle, which in turn cause areas of crust to move about.

> **a** Explain how the circulation of soup in the saucepan above can be used to represent convection in the Earth's mantle.

### Tectonic plates

In fact, the Earth's crust isn't a single continuous shell like an egg shell. It has been cracked and split into several huge pieces. We call these pieces **tectonic plates**. They are made up of the Earth's crust and the upper part of the mantle. These plates are slowly moved around by the convection currents within the mantle. This has been happening for billions of years.

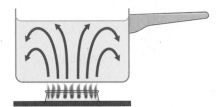

**Figure 1** Convection currents are caused when a fluid is heated. The warmer fluid is less dense than its cooler surroundings, so it rises.

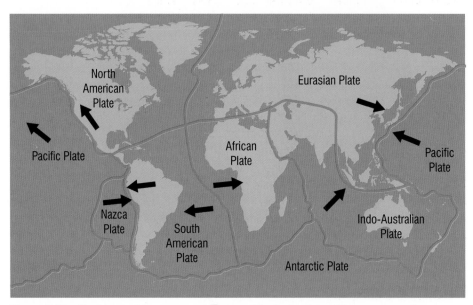

**Figure 2** The Earth's crust is composed of several tectonic plates. The arrows on the diagram show their direction of movement

> **b** How many tectonic plates are shown in the diagram above?
>
> **c** Which plates are moving toward each other, and which are slipping past each other?

### ??? Did you know ... ?

Tectonic plates also move away from each other, with material from the mantle rising up to fill the gap. In this way, the UK and US move about 2.5 cm further apart every year.

## Earthquakes and volcanoes

**Seismologists** and **volcanologists** study the places where these plates meet (**plate boundaries**). They do this because these places are often the sites of **earthquakes** and **volcanoes**.

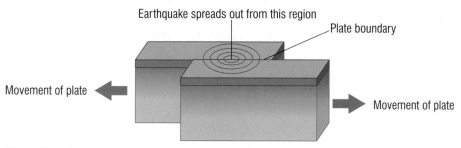

**Figure 3** Earthquakes are caused when plates grind against each other

Seismologists study earthquakes. Enormous strain builds up at the tectonic plate boundaries. When the force pushing the plates gets high enough, the plates can suddenly slip a short way past each other. This causes them to shudder as they grind against each other. Buildings nearby can collapse or be severely damaged. For example, in 2009, an estimated 230 000 people died when an earthquake hit Haiti.

Volcanologists study volcanoes. These occur when one plate slides under another. The plate being forced downward then starts to melt as it is pushed deeper into the mantle. The molten crust material is less dense than the solid crust above it. This allows it to rise back upwards, breaking through the surface to form a volcano.

**d** Why does molten crust material return to the Earth's surface?

Seismologists and volcanologists cannot stop earthquakes and volcanoes. However, they can help save lives by predicting when they will strike. A tool used by seismologists is the **seismometer**, which detects vibrations in the Earth's crust. By placing seismometers in different locations, they can build up a picture of how the crust is behaving. They can also detect tell-tale vibrations just before an earthquake. Volcanologists can use **GPS** (Global Positioning System) devices to detect whether a volcano is swelling up, which happens just before an eruption. This again gives some advance warning but accurate predictions are still very difficult to make.

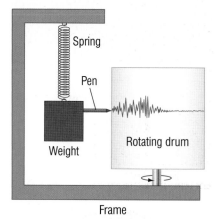

**Figure 4** Vibrations in the Earth's crust cause this simple seismometer to move. The weight and spring hold the pen still, so a pattern is drawn when the rotating drum moves.

### Summary questions

1 Copy and complete using the words below:

*toward   core   past   plates   mantle*

The heat of the Earth's ............ causes convection currents in the ............ . These currents move the tectonic ............ of the Earth's crust around. When tectonic plates move ............ each other, earthquakes can occur. When tectonic plates move ............ each other, volcanoes can be formed.

2 Describe how the movement of tectonic plates can cause earthquakes and volcanoes.

3 Evaluate the usefulness of the methods scientists use to predict earthquakes and volcano eruptions.

### Key points

- Tectonic plates are made up of the Earth's crust and upper part of its mantle.
- Heat from the Earth's core causes convection currents in the mantle.
- Convection currents in the mantle cause tectonic plates to move.
- Earthquakes and volcanoes occur mainly at boundaries between tectonic plates. Accurate predictions of events are very difficult to make.

## 1.6

# The Earth's changing atmosphere

## The Earth's atmosphere

The Earth's **atmosphere** was first formed billions of years ago. It was very different from the Earth's current atmosphere. Today, the gases surrounding us are mainly nitrogen with about one-fifth oxygen and trace amounts of other gases. The Earth's early atmosphere would not have been able to support life.

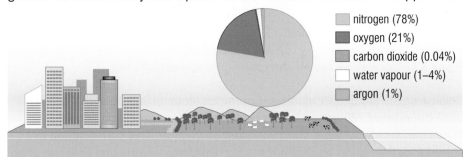

- nitrogen (78%)
- oxygen (21%)
- carbon dioxide (0.04%)
- water vapour (1–4%)
- argon (1%)

**Figure 1** The Earth's current atmosphere contains mainly nitrogen and oxygen, with small amounts of carbon dioxide and water vapour

The young Earth's first proper atmosphere was probably produced by the volcanoes that covered its surface. The gases volcanoes release into the atmosphere are mainly carbon dioxide and water. This is probably what the Earth's early atmosphere contained:

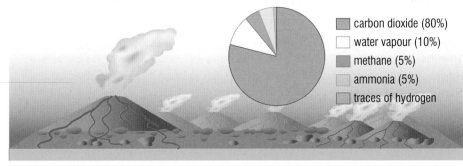

- carbon dioxide (80%)
- water vapour (10%)
- methane (5%)
- ammonia (5%)
- traces of hydrogen

**Figure 2** The Earth's first real atmosphere was made from gases released by volcanoes

## Photosynthesis

As the water vapour in the atmosphere cooled and condensed, the oceans started to form. Some single-celled organisms, such as algae living in the oceans, used the carbon dioxide in **photosynthesis**. During photosynthesis carbon dioxide and water form glucose for the organism to use as energy, and oxygen gas is given off. As the population of these microorganisms grew, they started to absorb more and more of the carbon dioxide in the atmosphere.

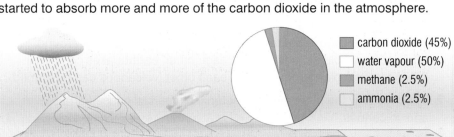

- carbon dioxide (45%)
- water vapour (50%)
- methane (2.5%)
- ammonia (2.5%)

**Figure 3** Microorganisms evolve in the first oceans

∞ links

*For more information on photosynthesis see 4.1 Biomass and food chains.*

The early microorganisms continued to grow, producing oxygen. The oxygen built up slowly. It reacted with methane, producing carbon dioxide and water. The oxygen also reacted with the ammonia, producing nitrogen and more water.

**a** Write word equations to show how oxygen removed ammonia and methane from the Earth's early atmosphere.

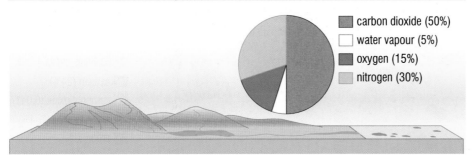

- carbon dioxide (50%)
- water vapour (5%)
- oxygen (15%)
- nitrogen (30%)

**Figure 4** Microorganisms continue to evolve. The resulting oxygen from photosynthesis has removed the methane and ammonia from the atmosphere.

Photosynthesis continued to reduce the amount of carbon dioxide in the atmosphere, and increase the amount of oxygen. Nitrogen continued to build up as oxygen reacted with ammonia. Nitrogen is unreactive, so it just stayed in the atmosphere. When larger plants evolved, these also carried out photosynthesis. These are believed to produce at least half of the oxygen there is in the atmosphere now.

**b** Write a word equation for photosynthesis.

*Higher*

Because humans weren't around billions of years ago, the description of how the atmosphere developed is theoretical. It is based on good evidence researched by geochemists, but it is only one of many theories. It also provides clues about how life began. In the 1950s, a pair of chemists, Miller and Urey, simulated Earth's early atmosphere. They fired electricity through a mixture of ammonia, methane, hydrogen and water. This produced amino acids, the building blocks of proteins essential to life.

## Summary questions

**1** Copy and complete using the words below. You may use a word more than once.

*oceans  oxygen  volcanoes  nitrogen  carbon  water  hydrogen*

The Earth's present atmosphere contains mainly ............. and ............., with small amounts of ............. dioxide, ............. and ............. vapour. The gases in the Earth's early atmosphere were produced by ............. They were mainly carbon dioxide and ............. vapour. Early microorganisms were able to evolve in the first ............. These converted carbon dioxide into ............. gas.

**2** Explain how the concentration of carbon dioxide in the atmosphere has changed in the past 4 billion years.

**3** Some of the carbon dioxide from the early atmosphere is locked up in fossil fuels. Suggest how this could have happened.

 **Examiner's tip**

It is important that you know the gases involved in photosynthesis. You may need to explain changes on a graph or table showing how these gases have changed during Earth's history.

**Did you know ... ?**

It's unlikely all the water on Earth came from volcanoes. One theory is that it arrived in **comets** that collided with the early Earth billions of years ago.

## Key points

- The Earth's early atmosphere was produced by volcanoes, and was mainly carbon dioxide, water vapour, methane, hydrogen and ammonia.

- As the water vapour cooled, the first oceans formed and early plants began to evolve.

- Early microorganisms used carbon dioxide in the atmosphere for photosynthesis. This made oxygen gas, which in turn converted ammonia into nitrogen.

# 1.7 Maintaining our atmosphere

- What is the natural greenhouse effect?
- How do greenhouse gases help keep temperatures stable and warm enough to support life?

The Earth's atmosphere is mainly made up of nitrogen and oxygen, along with a tiny amount of carbon dioxide and water vapour. Another gas present in the atmosphere is argon. The name argon comes from the Greek word for 'inactive' because it doesn't react. The other main gases, however, have important roles in supporting life on the Earth.

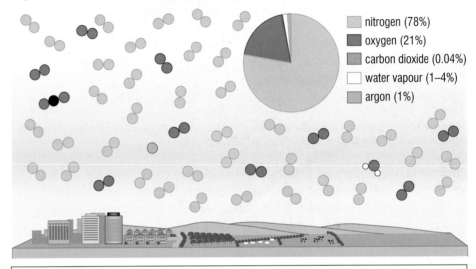

nitrogen (78%)
oxygen (21%)
carbon dioxide (0.04%)
water vapour (1–4%)
argon (1%)

**Nitrogen ($N_2$):** This gas makes up most of the atmosphere. It doesn't easily react with other chemicals. It can sometimes react with oxygen to form nitrates, which are absorbed by plants and used in making proteins.

**Water vapour ($H_2O$):** The amount of water vapour in the atmosphere varies a lot depending on where you are. It is usually between 1 and 4 per cent.

**Carbon dioxide ($CO_2$):** The tiny amount of carbon dioxide in the atmosphere is enough for all the world's plants to make the food they need to live. Through photosynthesis, plants convert carbon dioxide and water into glucose and oxygen.

**Oxygen ($O_2$):** the gas living organisms use in respiration. Not much life could exist here without it. Whenever we burn things, they react with oxygen and form oxides.

**Argon (Ar):** Trace gases like argon make up around 1 per cent of the Earth's atmosphere.

**Figure 1** The main gases found in the Earth's atmosphere

**a** Which gases in our atmosphere are used by living organisms?

## Greenhouse gases

**Greenhouse gases** have quite a bad reputation. The **greenhouse effect** is often in the news, along with terms such as **global warming** and **climate change**. What the papers often don't say is that without greenhouse gases, there would be very little life on the Earth.

As you know, life could not exist on the Earth without the Sun. The Sun provides light energy for plants to use during photosynthesis. It also provides energy which prevents the planet freezing. Energy is radiated by the Sun as **short-wave radiation** (mainly light energy), which warms the surface of the Earth. The warm Earth then radiates **long-wave radiation** back out into space. Greenhouse gases let short-wave radiation through, but they absorb long-wave (infrared) radiation. This 'traps' energy in the atmosphere, keeping the planet warm.

## Practical

### A greenhouse gas experiment

You can demonstrate the greenhouse effect in the lab. Fill three clear bottles with different gases: carbon dioxide, methane and air. Insert a thermometer through the cap of each bottle and leave them in bright sunlight. Record the temperature in each bottle at the start and after some time. Which bottles warm up fastest?

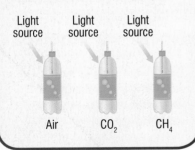

Light source    Light source    Light source

Air    $CO_2$    $CH_4$

**b** How do greenhouse gases raise the Earth's temperature?

This isn't as bad as it sounds. Without these gases, the Earth would be about 33 °C colder than it is now. Most of the planet would be encased in ice. However, humans are increasing the amount of greenhouse gases in the atmosphere. This is making the planet warmer and, scientists argue, might make our climate more extreme.

**links**

*For more information on greenhouse gases see 14.1 Producing greenhouse gases and 14.2 The effects of greenhouse gases.*

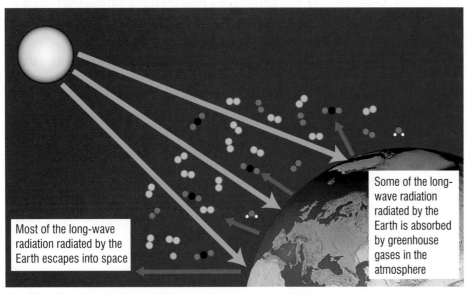

Most of the long-wave radiation radiated by the Earth escapes into space

Some of the long-wave radiation radiated by the Earth is absorbed by greenhouse gases in the atmosphere

**Figure 2** The Sun radiates short-wave radiation, which warms the Earth. Greenhouse gases stop the Earth radiating long-wave radiation back into space

**c** Why are greenhouse gases useful to have in the atmosphere?

**d** What problems are caused by too much greenhouse gas?

The main greenhouse gas is carbon dioxide. It is put into the atmosphere by the respiration of living things and by burning **fossil fuels**. Another major greenhouse gas is **methane**. Methane is one of the waste products of cattle and other livestock. Water vapour also acts as a greenhouse gas.

 **Did you know … ?**

Methane is produced by cattle and its warming effect is 23 times greater than that of carbon dioxide. However, there is much more carbon dioxide in the air than methane.

## Summary questions

**1** Copy and complete using the words below:

*oxygen methane greenhouse Sun carbon short-wave nitrogen long-wave water nitrates argon*

The Earth's atmosphere is 78 per cent ............., 21 per cent oxygen, and also contains ............, ............ dioxide and ............ vapour. Nitrogen can form ............ which are used by plants. ............ is used by most living organisms in respiration. The Earth is warmed by ............ radiation from the ............. The Earth radiates ............ radiation back into space, but ............ gases can absorb and trap it. Examples of these gases are carbon dioxide and ..............

**2** Make a table with the headings GAS, ROLE and PERCENTAGE. Use the information in Figure 1 to complete your table.

**3** Describe and explain two ways in which carbon dioxide supports life on Earth.

## Key points

● The Earth's atmosphere contains mainly nitrogen and oxygen, with small amounts of carbon dioxide, water vapour and argon.

● Short-wave radiation (light energy) from the Sun passes through the atmosphere to warm the Earth.

● Long-wave radiation emitted by the Earth can be absorbed by greenhouse gases. This natural greenhouse effect keeps the Earth warm enough to inhabit.

# Summary questions

1. Match up the following words with their descriptions:
   a Inner core     Liquid metal
   b Outer core     Slowly flowing rocks
   c Mantle        Solid rock
   d Crust         Solid metal

2. Explain why we need to use different telescopes to gain a full idea of what is in the universe, including space-based telescopes, land-based telescopes and telescopes sensitive to different types of electromagnetic radiation.

3. Explain what the red shift tells us about the universe. Describe the evidence for the Big Bang.

4. Describe how our model of the solar system has changed. How did telescopes help us develop new models?

5. Explain why we cannot send a space probe to the star nearest the Sun.

6. Explain why we are less likely to find life on Mars than to find life in another solar system.

7. Mars is our closest planet but it is further away from the Sun than we are. Suggest three features of Mars that would be different as a result of this.

8. Seismologists make measurements in the Earth's crust. What sort of measurements do they make, and what equipment do they use?

9. Look at the graph and then answer the questions.

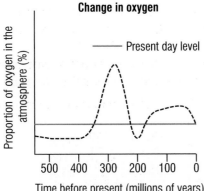

**Change in oxygen**

— Present day level

Proportion of oxygen in the atmosphere (%)

500  400  300  200  100  0
Time before present (millions of years)

   a How many years ago was the oxygen level the highest?
   b Suggest the year that plants first started growing on the Earth.
   c The present day level of oxygen is 21 per cent. Suggest the proportion of oxygen 200 million years ago.
   d How did plants affect the concentrations of oxygen and carbon dioxide in the Earth's early atmosphere?

10. What happened to the ammonia and methane believed to be in the Earth's early atmosphere?

11. Give **two** reasons why it is important for there to be carbon dioxide in our atmosphere.

12. Explain in steps how energy from the Earth's core can end up causing an earthquake. Use diagrams to help explain each step.

13. Global warming is sometimes described as carbon dioxide 'trapping' energy in the Earth's atmosphere. Explain in detail how this happens.

# AQA Examination-style questions

**1** Put the following in order of size starting with the smallest.

**A** Moon
**B** Sun
**C** Milky Way
**D** Earth
**E** Solar system (4)

**2** Choose words from the box to label the diagram of the Earth.

> Core    Crust    Atmosphere
> Mantle    Ozone

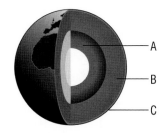

(3)

**3** Complete the table.

| Part of the early atmosphere | Formula |
|---|---|
|  | $H_2O$ |
| Carbon dioxide |  |
|  | $CH_4$ |
| Ammonia |  |

(4)

**4** The order of the planets in our solar system is given in the table below.

For each planet, choose the correct length of its year.

| Planet | Year length in Earth years |
|---|---|
| Mercury | 164.81 |
| Venus | 11.86 |
| Earth | 247.70 |
| Mars | 1.88 |
| Jupiter | 0.62 |
| Saturn | 29.456 |
| Uranus | 84.07 |
| Neptune | 0.24 |
| Pluto (a dwarf planet) | 1.00 |

(3)

**5** Match the type of wave each telescope uses to how the telescope may be used by astronomers.

| Type of wave received by telescope | Measuring temperature | Seeing through clouds and smoke | Looking at distant neutron stars |
|---|---|---|---|
| Gamma |  |  |  |
| Infrared |  |  |  |
| Microwave |  |  |  |

(2)

**6** Choose the correct word from each box to finish these sentences.

**a** ............ is a greenhouse gas.

> Argon    Carbon monoxide    Methane

(1)

**b** Greenhouse gases absorb more ............ radiation.

> short-wave    medium-wave    long-wave

(1)

**c** Therefore more heat is retained in the ............ .

> atmosphere    ground    sea

(1)

**7** *In this question you will be assessed on using good English, organising information clearly and using specialist terms where appropriate.*

Explain how scientists used line spectra to explain what is happening to the universe. (6)

**8** Telescopes can be used on Earth and in space.
**a** Suggest **two** benefits to astronomers who use telescopes on the surface of the Earth. (2)
**b** Suggest **two** reasons why other astronomers would prefer to use a telescope on a satellite in space. (2)

**9** The diagram shows the plate boundary between North America and Mexico.

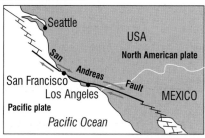

**a** Explain what happens inside the Earth so that earthquakes occur. (3)
**b** Suggest **two** reasons why there are a lot of earthquakes in Los Angeles. (1)
**c** Suggest another town that might suffer from a lot of earthquakes (1)

# 2.1 Building blocks of new products

## Learning objectives

- How do we classify materials?
- What is the difference between an element, a compound and a mixture?

**Materials scientists** work with thousands of different materials, each with its own properties. The materials in new products are chosen carefully so they have the right properties. To make things easier to manage, materials can be put into three groups: **elements**, **compounds** and **mixtures**.

## Elements

**Atoms** are the building blocks for all materials. Materials made of only one type of atom are called elements. The **periodic table** is a list of all the different elements.

Group numbers

| | | | | | | | | | | | | | | | | | 0 |
|---|---|---|---|---|---|---|---|---|---|---|---|---|---|---|---|---|---|
| | | | | | | | | | | | | | | | | | He 2 Helium |
| 1 | 2 | | H 1 Hydrogen | | | | | | | | | 3 | 4 | 5 | 6 | 7 | |
| Li 3 Lithium | Be 4 Beryllium | | | | | | | | | | | B 5 Boron | C 6 Carbon | N 7 Nitrogen | O 8 Oxygen | F 9 Fluorine | Ne 10 Neon |
| Na 11 Sodium | Mg 12 Magnesium | | | | | | | | | | | Al 13 Aluminium | Si 14 Silicon | P 15 Phosphorus | S 16 Sulfur | Cl 17 Chlorine | Ar 18 Argon |
| K 19 Potassium | Ca 20 Calcium | Sc 21 Scandium | Ti 22 Titanium | V 23 Vanadium | Cr 24 Chromium | Mn 25 Manganese | Fe 26 Iron | Co 27 Cobalt | Ni 28 Nickel | Cu 29 Copper | Zn 30 Zinc | Ga 31 Gallium | Ge 32 Germanium | As 33 Arsenic | Se 34 Selenium | Br 35 Bromine | Kr 36 Krypton |
| Rb 37 Rubidium | Sr 38 Strontium | Y 39 Yttrium | Zr 40 Zirconium | Nb 41 Niobium | Mo 42 Molybdenum | Tc 43 Technetium | Ru 44 Ruthenium | Rh 45 Rhodium | Pd 46 Palladium | Ag 47 Silver | Cd 48 Cadmium | In 49 Indium | Sn 50 Tin | Sb 51 Antimony | Te 52 Tellurium | I 53 Iodine | Xe 54 Xenon |
| Cs 55 Caesium | Ba 56 Barium | Lanthanides | Hf 72 Hafnium | Ta 73 Tantalum | W 74 Tungsten | Re 75 Rhenium | Os 76 Osmium | Ir 77 Iridium | Pt 78 Platinum | Au 79 Gold | Hg 80 Mercury | Tl 81 Thallium | Pb 82 Lead | Bi 83 Bismuth | Po 84 Polonium | At 85 Astatine | Rn 86 Radon |
| Fr 87 Francium | Ra 88 Radium | Actinides | | | | | | | | | | | | | | | |

The transition metals

The alkali metals    The alkaline earth metals

The halogens    The noble gases

**Figure 1** The periodic table of elements

Look at the periodic table in Figure 1. You can probably see some materials you have already heard of, such as iron, oxygen, gold and tin. Pure elements are rarely used on their own. It is often more useful to combine elements into compounds. This makes it possible to create products with a wider range of properties.

**a** Try to name the elements in the pictures below.

### Did you know ...?

The first periodic table was created in 1869 by a Russian chemist called Dmitri Mendeleev.

**Figure 2** Each of these contains only one type of atom. They are all elements.

## Compounds

Most chemicals you have used in science lessons are compounds. Compounds all contain more than one type of atom.

The chemical bonds in compounds hold the atoms together tightly. This makes it difficult to separate a compound back into its elements.

Combining elements together in compounds can be useful. The new material will have completely different properties from the elements it is made of. For example, kitchen salt is a compound made of sodium and chlorine. Pure sodium can easily burn through your skin, and chlorine is a poisonous gas. However, sodium chloride is used in cooking every day.

**b** Which elements are in the two compounds shown in Figure 3?

**Figure 3** Salt (NaCl) and water ($H_2O$) are both compounds

## Mixtures

Most of the products and materials you see every day are mixtures. There are two key things you need to know about mixtures:

- they contain more than one type of substance
- the different substances are not bonded to each other.

Because the different substances are not bonded together, they are easier to separate. By combining substances to make mixtures, materials scientists can fine tune their properties. For example, steel is a mixture of iron and carbon. Iron is quite strong, but adding just 1 per cent carbon makes it even stronger.

**c** What is the difference between a mixture and a compound?

**Figure 4** There are lots of different compounds in these mixtures

---

### Summary questions

1 Copy and complete using these words:

*compounds   atoms   contain   substances   elements*

All substances are made of ............ Substances made up of only one type of atom are called ............ Substances which ............ more than one type of element bonded together are called ............ In mixtures, different ............ are not bonded to each other.

2 Look at the following diagrams. Use the descriptions on this spread to decide which is an element, which is a compound and which is a mixture.

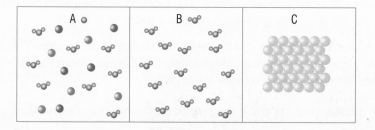

### Key points

- An element is made of only one type of atom.

- Compounds contain more than one type of atom, bonded together.

- Mixtures contain more than one type of element or compound, not bonded together.

<table>
<tr><td>

**2.2**

</td><td>

# Inside atoms

</td></tr>
</table>

**Figure 1** The Large Hadron Collider at CERN will help answer questions about the universe

Scientists need to understand how the atoms in a substance behave. This makes it easier to control the properties of a new material. To understand atoms, scientists needed to explain what they are made of.

Atoms are very, very small. In fact, if you laid about 1 billion in a line they would be as long as the full stop at the end of this sentence.

About 100 years ago, scientists discovered that the atom is made up of even smaller particles. These are called subatomic particles. There are three main types of subatomic particle you need to learn about: protons, neutrons and electrons.

## Atomic structure

By knowing the structure of its atoms, you can predict how an element will behave. For example, will it conduct energy? Or, will it react with other substances? Look at the picture below, it shows two simple atoms. You can see they are both made from the same three types of particle.

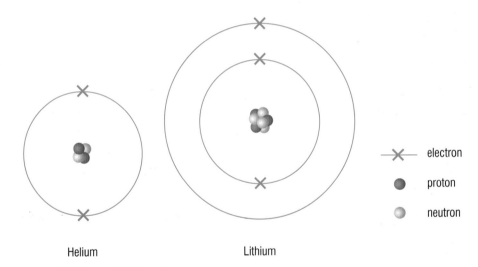

Helium          Lithium

**Figure 2** Helium and lithium atoms

There are two types of particle in the centre or **nucleus** of the atom. These are called **protons** and **neutrons**. The particles orbiting around the outside are called **electrons**.

The particles have different charges and masses. The protons are positive. The neutrons are neutral and the electrons are negative.

The nucleus is small and very dense. The protons and neutrons are the heavy particles in an atom. We give them a relative mass of 1.

The smaller electrons have hardly any mass and orbit the nucleus. We call the area that the electrons orbit an electron shell. Because electrons have hardly any mass they are given a relative mass of 0.

a  How many protons are in the helium atom shown in Figure 2?

b  How many neutrons are in the lithium atom shown in Figure 2?

## Atomic number and mass number

The helium and lithium atoms are different elements because they have different numbers of protons. The number of protons an atom has tells you what element that atom is. So if an atom has three protons, it has to be a lithium atom.

An atom contains the same number of protons (positive charge) and electrons (negative charge), so lithium has three electrons.

If you look at a periodic table, you will see that next to every element there are often two numbers. The smaller number tells you the number of protons in an atom of that element. This is called the **atomic number**, for example:

$$_6C \quad _2He \quad _{10}Ne$$

**c** Use the periodic table to find the atomic numbers of iron, chlorine, sodium and gold.

The top number is called the **mass number**. This tells you the total number of protons plus neutrons in an atom. A lithium atom has three protons and four neutrons. So its mass number is 7.

| Particle | Where is it? | Charge | Mass |
|---|---|---|---|
| Proton | In the nucleus | Positive | 1 |
| Neutron | In the nucleus | Neutral | 1 |
| Electron | In shells outside the nucleus | Negative | Almost zero |

So to recap:

- All atoms of the **same** element have the **same** number of protons.
- The **atomic number** of an element is the number of **protons** its atoms contain.
- The number of electrons in an atom is **equal** to the number of protons.
- The **mass number** tells us the number of **protons plus neutrons**.
- If you want to work out the number of neutrons in an atom, just **subtract** the atomic number from the mass number.

**d** Use the periodic table to calculate the number of neutrons in a potassium atom with a mass number of 39.

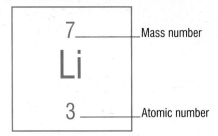

**Figure 3** The atomic and mass number of lithium. We show this as $_3^7Li$.

### AQA Examiner's tip

You may be asked to draw and label a diagram of the atom when you are given just the atomic number and mass number, so make sure you know how to do this.

### ??? Did you know ...?

Atoms may be small, but they can release huge amounts of energy. Some elements, such as uranium, have unstable nuclei. When a uranium nucleus is hit by a neutron it breaks apart, releasing energy and several neutrons. Each neutron then hits another uranium atom. This process is called a chain reaction.

If this is controlled, you can generate nuclear power. An uncontrolled nuclear chain reaction takes place in a nuclear bomb.

### Summary questions

**1** Copy and complete using these words:

*nucleus   three   positive   one   negative*
*outside   protons   atomic*

Atoms contain ............ types of particles. Neutrons are neutral and are found in the ............ Protons, which have a ............ charge, are also found in the nucleus. Protons and neutrons have a relative mass of ............ Electrons have a ............ charge and are located ............ the nucleus in shells. All atoms of the same element have the same number of ............ This is shown on the periodic table by the ............ number.

**2 a** Why is the nucleus the densest part of an atom?
  **b** An atom contains nine protons. What type of atom is it?
  **c** All atoms are neutral in charge. A particular atom contains five protons, how many electrons will it have?

### Key points

- Protons are positive, neutrons are neutral and electrons are negative.
- The number of protons in an atom equals the number of electrons in that atom.
- Protons and neutrons are in the nucleus of an atom and have a relative mass of 1.
- Electrons have a very tiny mass. They orbit the nucleus.

# 2.3 Different types of particles

## Learning objectives

- What is the difference between atoms, molecules and ions?
- Why are ions charged?

## Atoms and molecules

Incredibly, there are only 92 types of **atoms** that exist naturally. The other 20 or so have been created by humans. All the types of atom are listed in the periodic table.

Everything around you, even your own body, is made up from these atoms. We can use symbols and pictures to represent atoms.

**a** Look at the diagram below. Which three elements are shown?

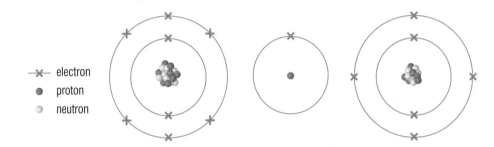

- ✕ electron
- ● proton
- ○ neutron

**Figure 1** Some examples of atoms. The first electron shell (nearest the nucleus) can hold 2 electrons and the second shell can hold 8 electrons. (The third electron shell also holds 8 electrons before the fourth shell starts to fill.)

Atoms are the building blocks for all materials. They can join together with chemical bonds to form **molecules**. The atoms in a molecule of an element are the same, such as the oxygen molecules in the air. Or they can be made up from different types of atom, such as water or carbon dioxide. Molecules made up from different types of atom are compounds.

- A molecule is two or more atoms that are chemically joined.
- A compound is made by two or more **different** types of atoms that are chemically joined.

$$O=O \qquad N\equiv N \qquad O=C=O$$

Oxygen      Nitrogen      Carbon dioxide

$$H-\underset{\underset{H}{|}}{\overset{\overset{H}{|}}{C}}-H$$

Methane

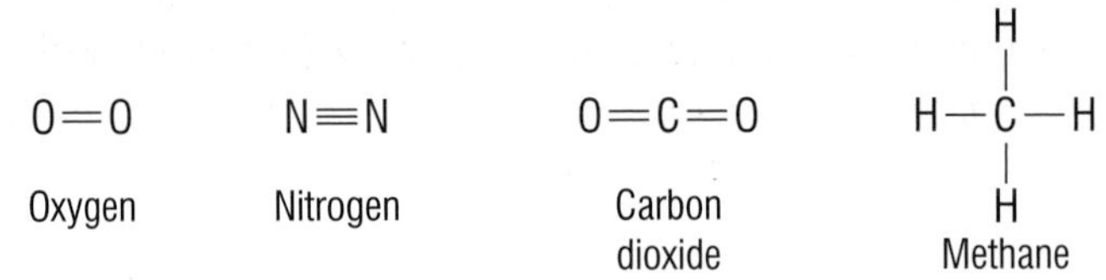

**Figure 2** Some examples of molecules

**b** Look at the molecules in Figure 2. Which are molecules of an element, and which are molecules of a compound?

Look around you. How many different materials can you see? All of these are made from atoms (usually in molecules). Some compounds have been discovered by accident, such as the sweetener aspartame. Some elements were discovered by accident, too. Phosphorus was accidentally discovered by a scientist boiling urine. It is now used to make match heads.

## Ions

Sometimes atoms can get a positive or negative charge. This is because they can gain or lose electrons. Remember, an electron has a negative charge, so:

- if an atom gains an electron it will become *negatively* charged by 1
- if an atom loses an electron it will become *positively* charged by 1.

We call these charged particles **ions**.

**c** If an atom gained two electrons, what would the charge on the ion be?

**d** If an atom lost three electrons, what would the charge on the ion be?

Some ions produce great colours when put in a Bunsen flame. If you needed to identify an unknown substance you could use a flame test.

Pyrotechnicians (fireworks scientists) use their knowledge of ions to choose the colours of fireworks.

**Figure 4** This firework colour is made using potassium ions, $K^+$

---

### Practical

#### Flame tests

To perform a flame test you need a mounted wire loop and a Bunsen burner. Clean the wire loop using the hottest flame of the Bunsen. Dip the loop in the substance you want to test and hold it over a blue flame. The flame will change colour. Different metal ions will produce different coloured flames.

**Safety:** Wear eye protection.

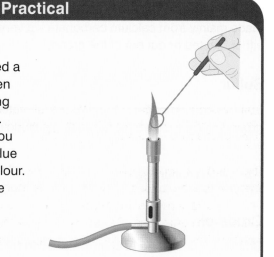

**Figure 3** Performing a flame test

---

An understanding of both ions and molecules is important when making products. Toy 'slime' can be made by adding borax (used in laundry powder) to PVA glue. The ions produced by the borax cause the molecules in the PVA to join together. The result is sticky slime.

**Figure 5** Making slime

### Summary questions

1 Copy and complete using these words:

*joined   ion   two   atoms*

Everything is made up of small particles called ............ A molecule is ............ or more atoms that are chemically ............ If an atom or a molecule becomes charged it is called an ............

2 Why is water a compound?

3 When an atom loses one of its electrons, does it become negatively or positively charged?

### Key points

- Atoms are the tiny particles which are the building blocks of all materials.

- A molecule is two or more atoms that are chemically bonded.

- If an atom or molecule becomes charged (by gaining or losing electrons), it forms an ion.

# 2.4 Making products with materials from the Earth

## Learning objectives

- Which raw materials can we use straight from the Earth?
- Which raw materials do we have to separate and refine before use?
- How do we separate mixtures like crude oil and rock salt?

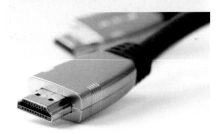

**Figure 1** As well as being decorative, gold is used in electronics

**Figure 2** Marble in the construction industry

### links

*For more information on limestone see 7.1 Limestone as a building material.*

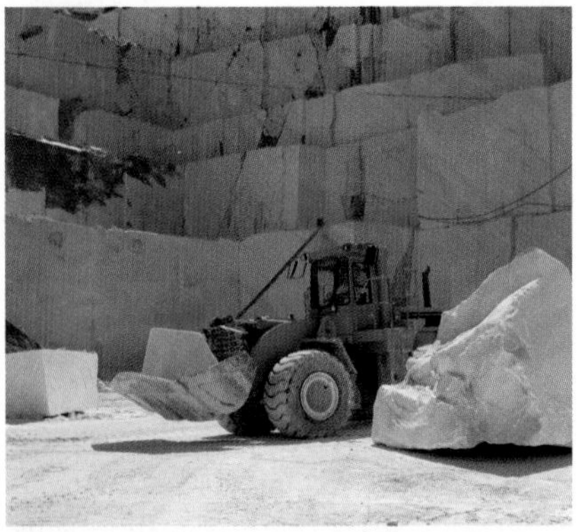

We all depend on the materials we can find in our environment. However, some materials are harder to get at than others.

## Gold

Since ancient times, gold has always been highly valued because of its attractive appearance. Gold does not easily react with other substances. This is why it stays shiny. This also makes it useful in electronics as its surface doesn't corrode. So it makes a good contact material for connections in microelectronic circuits.

**a** Is gold an element, a compound or a mixture?

## Limestone

Limestone and marble are very important construction materials. Both are made mainly from calcium carbonate – a very useful compound. They can be either blasted or cut out of the ground.

## Sulfur

Sulfur is an element. It is another useful material taken straight from the ground. It is important for producing sulfuric acid which makes fertilisers. We also use sulfuric acid in car batteries.

**b** Why is sulfur useful?

## Crude oil

Crude oil is a mixture of chemicals called **hydrocarbons**. A hydrocarbon is a compound containing *only* hydrogen and carbon.

**Petrol** and **natural gas** are important examples of hydrocarbon fuels. We also use hydrocarbons to make plastics, paints and some medicines. We can separate the hydrocarbons in crude oil because they have different boiling points.

There are patterns in how hydrocarbons of different sizes appear and behave. This comes in handy for separating them when crude oil is processed.

| Small hydrocarbons |
|---|
| Light in colour |
| Low boiling point |
| Catch fire easily |
| Thin and runny |

| Large hydrocarbons |
|---|
| Dark in colour |
| High boiling point |
| Don't catch fire easily |
| Thick and viscous |

Separating crude oil in this way is called **fractional distillation**. Crude oil is separated in enormous towers called fractionating columns. Its hot vapour enters the column. The tower is cooler higher up. As the hydrocarbons cool, fractions (mixtures of hydrocarbons with similar boiling points) of the vapour condense back into liquids at different heights. This is where the process gets its name from.

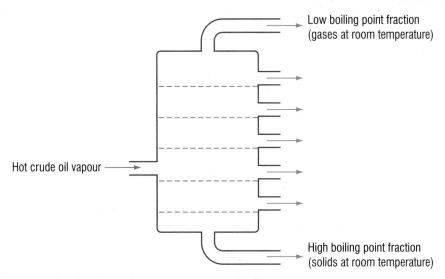

Low boiling point fraction
(gases at room temperature)

Hot crude oil vapour →

High boiling point fraction
(solids at room temperature)

**Figure 3** A fractionating column to separate crude oil

**Figure 4** Fractional distillation in industry

## Salt from rock salt

We mine huge amounts of salt from the ground. This type of salt is called rock salt.

Rock salt is a mixture of salt (sodium chloride) and rock. Rock salt can be easily separated by adding water to the mixture. The salt dissolves. Then we can remove the bits of insoluble rock by filtering. The salt is separated from its solution by evaporating off the water.

**Figure 5** Walls of salt in a salt mine

### Summary questions

1 **a** Name **one** material we use straight from the Earth and one mixture we need to separate.
  **b** Describe **two** uses for each raw material you named in **a**.

2 Why is gold often used to plate contacts in electronic devices?

3 How are we able to separate the hydrocarbons in crude oil?

4 Describe how you would get a pure sample of salt from rock salt in your school laboratory.

### Key points

- Gold, marble, limestone and sulfur can all be used straight from the ground.

- Crude oil is a mixture of hydrocarbons. We can separate hydrocarbons by fractional distillation.

- Pure salt can be extracted from rock salt by adding water and filtering out the rock. We can then get the salt from its solution by evaporating off the water.

## 2.5 Extracting metals for construction

### Learning objectives

- What is an ore?
- Which common metals do we get from ores?
- How can we use reducing agents to get metals from ores?

**Figure 1** Mining iron ore

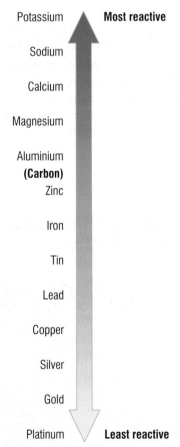

The reactivity series

Potassium — **Most reactive**

Sodium

Calcium

Magnesium

Aluminium

(Carbon)

Zinc

Iron

Tin

Lead

Copper

Silver

Gold

Platinum — **Least reactive**

**Figure 2** This reactivity series shows how reactive each element is compared to the other elements

Metals are vital to make products. Even products that don't contain metals are often made by machines that do.

Most metals exist in the Earth's crust as materials called **ores**. An ore is a rock containing a useful amount of metal. Ores contain enough metal to make it worth spending the money to extract it. We need to separate the metal from the other elements found in its compound in the ore.

Many ores contain oxides of the metal. These are compounds of the metal plus oxygen. We can remove oxygen from a compound with a **reducing agent**.

> **a** What is an ore?

An element used as a reducing agent has to be **more reactive** than the metal in the oxide. Being more reactive means it is able to 'take' the oxygen away from the metal. If this happens, we say that the metal oxide has been **reduced**.

### What will reduce what?

The **reactivity series** is used to work out which reducing agent could be used with which ore. A reactivity series is just a list of elements in order of how reactive they are (see Figure 2). Carbon is more reactive than iron or lead. This means we can use carbon or carbon monoxide as a reducing agent for iron and lead ores. In fact, any metal less reactive than carbon can be extracted in this way.

> **b** What is a reducing agent?
>
> **c** Name **two** metals whose ores could **not** be reduced by carbon. (Hint: look at the reactivity series.)

### Producing iron from iron ore

Iron is a very important metal. We use it to make steel. Most iron comes from an ore called haematite. This contains iron oxide ($Fe_2O_3$).

Haematite is crushed and put into a huge tower called a **blast furnace**. Carbon (in the form of coke made from coal) is also added. Air is blasted through the furnace at high temperature. The following chemical reaction happens:

$$C \quad + \quad O_2 \quad \longrightarrow \quad CO_2$$

carbon      oxygen      carbon dioxide

Then the carbon dioxide reacts with more carbon (coke) to produce carbon monoxide:

$$C \quad + \quad CO_2 \quad \longrightarrow \quad 2CO$$

carbon   carbon dioxide   carbon monoxide

The carbon monoxide then reduces the iron oxide:

$$3CO \quad + \quad Fe_2O_3 \quad \longrightarrow \quad 3CO_2 \quad + \quad 2Fe$$

carbon monoxide   iron oxide   carbon dioxide   iron

In the process of reducing iron oxide, carbon monoxide is itself **oxidised**. It has oxygen added to it to form carbon dioxide.

## Extracting lead from its ore

Lead is used in car batteries and roofing for buildings. Because lead is less reactive than iron, it is easier to extract than iron. It is easier to reduce lead oxide than iron oxide.

Carbon (again as coke) is added to lead oxide. This reduces the lead and makes carbon monoxide:

$$PbO + C \longrightarrow Pb + CO$$
lead oxide · carbon · lead · carbon monoxide

The carbon monoxide also reduces some of the lead oxide:

$$PbO + CO \longrightarrow Pb + CO_2$$
lead oxide · carbon monoxide · lead · carbon dioxide

## Extracting more reactive metals

Metals that are more reactive than carbon need to be extracted another way. These are extracted using a process called **electrolysis**. Electrolysis uses electricity to separate compounds. Aluminium is extracted from its ore (bauxite) in this way.

First of all aluminium oxide is separated from the bauxite and melted in a large electrolytic cell. Then an electric current is passed through it. The aluminium oxide is melted to allow the movement of ions in the molten mixture. Because aluminium forms positive ions, it is attracted to the negative part of the cell. The pure molten aluminium can then be poured out of the cell.

**Figure 3** Working at a blast furnace

⚭ links
*For more information on symbol equations see 2.9 Using equations.*

**AQA Examiner's tip**

At Higher Tier you are expected to be able to write balanced symbol equations for the reactions involved in separating metals from their ores.

⚭ links
*For more information on electrolysis see 12.1 Electrolysis and 12.2 Electroplating.*

**Figure 4** Electrolytic cells at an aluminium plant

### Summary questions

1 What is iron's main ore called?

2 What are the **two** reducing agents mentioned on these two pages?

3 Why could you not use carbon to reduce aluminium oxide into aluminium?

### Key points

- Reducing agents remove oxygen from ores.
- An element used as a reducing agent has to be more reactive than the metal in the oxide.
- Iron oxide and lead oxide are reduced by carbon and carbon monoxide.

## 2.6

# Products from the atmosphere

### Learning objectives

- How can we separate and collect gases in the atmosphere? [H]
- How do we use gases we obtain from the atmosphere? [H]

### ??? Did you know ...?

Cryonics is the science of deep freezing organisms using liquid nitrogen and other coolants. Hundreds of people have had their body frozen after death. They hope to be revived in the distant future. They believe that by then medical science will be advanced enough to cure the cause of their death.

**Figure 2** A container of liquid nitrogen

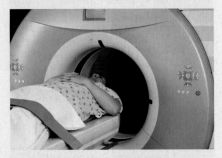

**Figure 3** Liquid helium cools the superconducting magnets in this MRI scanner

Most of the products we use from day to day are solids and liquids. However, without the gases we take from the atmosphere, many products couldn't be made at all. These gases are used when making light bulbs, neon signs, fertilisers, or when freezing things. Some of the gases can be used directly, for instance argon. Some gases, such as nitrogen, are used to make other products.

**a** Name **two** gases obtained from the Earth's atmosphere.

### Extracting gases from the air

Air is a mixture of mainly nitrogen and oxygen. Air also contains some 'trace' gases, such as argon, carbon dioxide and helium. To use gases from the air, they must first be separated. To do this, air is compressed and cooled. This changes it into a liquid at a very low temperature. The liquefied gases can be separated by fractional distillation in the same way as crude oil. Figure 1 shows how nitrogen and oxygen can be fractionally distilled using this method. The bottom part of the column is not as cold as the top part. Because nitrogen has a lower boiling point than oxygen, it rises to the top of the column as a gas.

### Using nitrogen

Nitrogen ($N_2$) is the most used gas from the atmosphere. Most of it is used to produce **ammonia** ($NH_3$). Ammonia is then used to make **fertilisers**, cleaning fluids and nitric acid.

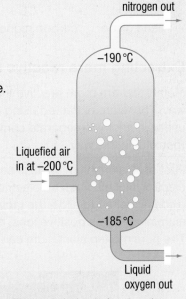

**Figure 1** A simple fractionating column separating oxygen and nitrogen

Pure nitrogen is useful by itself, either as a gas or a liquid. Because nitrogen is usually unreactive, it is used in the food industry as a **preservative**. For example, nitrogen is pumped into crisp packets before they are sealed. This replaces any oxygen and prevents **microbes** from growing. Liquid nitrogen is incredibly cold at $-196\,°C$, and can be used for freezing things very quickly. This method is used in medicine to keep samples of cells (such as sperms or eggs) for a long time.

**b** Describe **two** ways nitrogen can be used.

### Using helium

Helium is almost totally unreactive and much lighter than air. This makes it very useful. It can be extracted from air in the same way as nitrogen, but most is obtained from natural gas from oil fields. Helium is used in balloons and airships because of its very low density. Like liquid nitrogen, liquid helium is very cold, existing at $-269\,°C$. It is used to cool **superconductors** used in **MRI scanners**.

**c** Describe how liquid helium can be used.

## Using argon

Like helium, argon is almost totally unreactive. It also makes up nearly 1 per cent of the Earth's atmosphere; much more than helium. The largest domestic use for argon is lighting. **Filament (incandescent) light bulbs** are filled with argon so that the thin metal filament glows but doesn't burn and snap.

Argon produces light when an electrical current is passed through it. Because of this, it is used in filament and electric discharge tubes. It is also used in medical **lasers** for eye surgery and to make 'plasma' lighting.

### Did you know … ?

Helium is actually the second most common element in the universe, despite being quite rare on the Earth.

**Figure 4** Filament (incandescent) light bulbs are filled with argon so the filament doesn't burn out

**Figure 5** Argon glows when electricity passes through it

**d** Why is argon used in filament light bulbs?

Here is a summary of the different gases extracted from the atmosphere, and some of their uses:

| Gas | Uses |
| --- | --- |
| Nitrogen | Making ammonia, freezing agent, preservative |
| Helium | Balloons, airships, coolant for superconductors |
| Argon | Filament light bulbs, electric discharge tubes, lasers |

### links

*There is more on the gases in the atmosphere in 1.6 The Earth's changing atmosphere and 1.7 Maintaining our atmosphere.*

### Summary questions

**1** Copy and complete using these words:

*nitrogen   airships   coolant   bulbs   discharge fertilisers   preservative   dense   atmosphere*

We use gases taken from the Earth's ............ to make new products. ............ is used to make ammonia, which is used in making ............ Nitrogen is also used as a ............ and a ............ Helium is used in balloons and ............ because it is much less ............ than air. Argon is used in filament light ............ , electric ............ tubes and lasers.

**2** Why is liquid helium more expensive than liquid argon?

**3** Why can nitrogen be used to produce other chemicals, but not argon?

### Key points

- Gases are extracted from the atmosphere by liquefying air and separating the different liquids by fractional distillation. [H]

- Argon is used mainly in lighting. [H]

- Helium is used for cooling and for filling balloons and airships. [H]

- Nitrogen is mainly used to make ammonia. [H]

# 2.7 Exploiting the Earth's resources

### Learning objectives

- How do materials taken from the Earth benefit society?
- What is the environmental impact of mining and quarrying?
- How can mining be managed sustainably?
- How can phytomining help clean up the mining process?

Mining and quarrying affect all of us. Every day we rely on products of materials taken from the Earth. We use electronics, travel by cars, buses and trains, use plastic products and build buildings of stone and rock.

We are constantly using other mined materials too. Our homes are made from brick, concrete and steel. We put salt from salt mines on our food. We use plastics made from oil that was extracted from the ground.

> **a** Name **three** materials extracted from the Earth and describe what they are used for.

**Figure 1** We benefit from products made from the Earth's resources every day but often forget the activities that make them

As well as the consumer benefits, society benefits from mining in other ways. More than 100 000 people in the UK alone work in the mining and quarrying industries. This is a lot when you consider that most materials in the UK are imported.

## What are the problems with taking materials from the ground?

Fossil fuels, stone and metal ores are just some of the materials we mine, dig and pump out of the Earth. Environmentalists argue that mining activity is harming the environment. Mines and quarries damage the appearance of the landscape and destroy wildlife habitats. Then there is the noise and dust that affects the nearby area. Also, mining activity increases traffic near the mine. This leads to more air pollution.

> **b** List all the ways mining can harm the environment. Put your ideas into three groups: land, water and air.

Air pollution isn't just caused by the dust and traffic. Processing ores pollutes the air as well. Metals such as iron and lead produce carbon dioxide when they are purified. Lead, in particular, is very poisonous. Sulfur dioxide gas is also produced when metal sulfide ores are heated. This causes acid rain.

**Figure 2** Transporting mined materials

In some mines, rain can cause toxic chemicals to wash into rivers and lakes. This causes even more environmental damage.

There can be social drawbacks with mines as well. The material being mined runs out, eventually. This means that mines don't stay open forever. This can cause problems in communities relying on mines for employment.

## Managing resources responsibly

The Planning and Compulsory Purchase Act 2004 set out new rules for developing land into mines and quarries. Any development plans now need to be assessed for **sustainability**. Sustainable development is using resources in a way that finds a balance between human and environmental needs.

Sites that will cause the least environmental damage are preferred. Plans are discussed with all the people they will affect before a mine is started. Also, the after-effects of the mine on the community are considered.

## Issues with taking materials from the atmosphere

Like any other industrial process, extracting gases from the atmosphere requires a lot of energy. The more gas is needed, the more energy must be used. Burning fossil fuels is still the most common way to generate electricity. Extracting gases from the atmosphere contributes to the problems of burning fossil fuels. Also, gas generation plants take up land, and some of the products, such as ammonia, can be harmful.

Extracting and using gases from the atmosphere can have benefits. Tens of thousands of people are employed directly working in gas plants. Air is readily available, so there is no need to mine, drill or quarry. The gases being used can easily be recycled back into the air.

## Phytomining

Another way to reduce the impact of making products is to find useful ways to remove toxic waste from the environment. Some plants are able to absorb large amounts of harmful materials from the soil and store them. This process is called **phytomining**. It can be used to improve the quality of soil around coal and metal mines. For example, a South African plant called *Berkheya coddii* can safely absorb nickel from the ground. The plants can then be harvested and even used as **bio-ores**, reducing the need for mining.

### Key points

- We benefit from the products, services and jobs provided by taking materials from the Earth.

- The environment can be damaged by mining operations.

- Managing mining sustainably reduces long-term environmental damage.

- Plants can be used to phytomine areas, absorbing potentially toxic minerals from soil.

### Summary questions

1 What are:
   a the potential social benefits of opening a mine?
   b the social costs of opening a mine?

2 Describe three ways mining can damage the environment.

3 How can we manage mining sustainably?

4 Give two reasons why extracting iron ore damages the environment more than extracting nitrogen.

5 Describe two ways in which phytomining reduces harm to the environment.

# Changing materials to make new products

## Learning objectives

- How can materials be changed?
- What happens to the particles during a chemical reaction?
- How do manufacturers decide how much material to use in large-scale reactions?

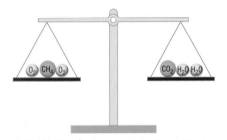

**Figure 1** What goes in must come out – the mass of products always equals the mass of reactants

**Figure 2** A demonstration of the thermite reaction, which can be be used to weld rails together

Most products we use start off as raw materials extracted from the Earth. In order to make these raw materials more useful, many need to be chemically changed. We call the process of chemically changing something a chemical reaction.

Sometimes, the ability to react is what makes a material useful. For example, antacids such as magnesium hydroxide are used to settle indigestion caused by excess acid in the stomach.

Chemists and materials scientists work with chemical reactions every day. They need to know how much useful material (**product**) they will produce. They also need to work out how much raw material (**reactant**) they will need to make it.

## The conservation of mass

Whenever chemicals react together, the total mass stays the same. This means the products always contain the same number of atoms as the reactants. Scientists call this the Law of the Conservation of Mass. This is because matter is conserved; it is not lost or gained. Because of this it is straightforward to work out how much product will be made.

### Practical

#### Conserving mass

You can use the reaction between an acid and a carbonate (marble chips) to prove that mass is conserved during a chemical reaction. Measure the mass of the acid and marble chips (as well as the conical flask). React the acid with the marble chips on top of the balance in the conical flask. Record any change in mass as the reaction progresses. You should notice that a gas escapes.

Repeat the experiment but this time with a balloon stretched over the top of the flask to stop gas escaping. Remember to measure the mass of the marble chips, acid, conical flask and balloon. What happens to the mass this time?

**a** What is the law of the Conservation of Mass?

## Thermite

The thermite reaction is used on railways to join rails together. Aluminium powder is heated up with iron oxide. Because aluminium is more reactive than iron, it 'takes' the oxygen. So the aluminium reduces the iron oxide to iron. We call this type of reaction a **displacement reaction**. Because the iron formed is molten, it joins the rails together as it cools:

aluminium + iron oxide ⟶ iron + aluminium oxide

$$2\ Al \quad + \quad Fe_2O_3 \quad \longrightarrow \quad 2\ Fe \quad + \quad Al_2O_3$$

We know that the total mass of aluminium and iron oxide will be the same as the mass of iron and aluminium oxide that is made. We can also calculate the exact proportions of aluminium and iron oxide needed for the reaction so that there is no waste.

**b** Why is aluminium able to reduce iron oxide to iron?

## Making calculations

An understanding of the conservation of mass helps manufacturers decide how much material to use. An example of this is in the production of zinc oxide.

### Zinc oxide

Zinc oxide is a very useful product. It is used in antiseptics, sunscreen, calamine lotion and shampoo. Zinc oxide is made by reacting zinc with oxygen:

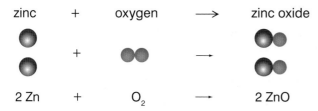

| zinc | + | oxygen | $\longrightarrow$ | zinc oxide |
| 2 Zn | + | $O_2$ | $\longrightarrow$ | 2 ZnO |

**Figure 3** Zinc oxide is used in sunscreen

If we know how much zinc and oxygen we use, we can calculate how much zinc oxide will be made. For example, if 65 g of zinc and 16 g of oxygen are used, the product must be 81 g of zinc oxide. This works both ways: if we have made 81 g of zinc oxide starting with 65 g of zinc, we must have used 16 g of oxygen. This can be scaled up or down, so if we had 32.5 tonnes of zinc and 8 tonnes of oxygen we would expect to make 40.5 tonnes of zinc oxide.

**c** If 162 g of zinc oxide was made using 32 g of oxygen, how much zinc must have been used?

## 2.9 Using equations

### Learning objectives

- How do manufacturers decide the amount of raw materials they need?
- How can we balance chemical equations? [H]

Choosing the right amount of materials to react with each other is very important. If the wrong amounts are used, then some reactants may be wasted. Waste means lost money to manufacturers and could also cause damage to the environment.

### Making fertiliser

If manufacturers understand a chemical reaction well they can choose the right quantities of reactants. The best way to start is with a word or symbol equation.

Ammonium nitrate is used as a fertiliser. It is an important chemical because it helps crops grow. It has the formula $NH_4NO_3$. The formula tells us it contains nitrogen, hydrogen and oxygen. The formula also tells us there are two nitrogen atoms, four hydrogen atoms and three oxygen atoms.

**a** How many different elements are present in ammonium nitrate?

**b** How many atoms are there in ammonium nitrate?

Ammonium nitrate is made by reacting ammonia ($NH_3$) with nitric acid ($HNO_3$). We can show this with the equation:

$$NH_3 \quad + \quad HNO_3 \quad \longrightarrow \quad NH_4NO_3$$

ammonia + nitric acid ⟶ ammonium nitrate

The symbol equation helps us see which atoms come from which reactant. We can also count how many atoms there are in the reactants and products. This helps us check whether the equation is '**balanced**'. For this reaction, the equation is already balanced. There are two N, four H and three O atoms on either side of the equation.

### Making ammonia

Making ammonia is an example of a reaction that has an equation that needs balancing. To produce ammonia ($NH_3$), hydrogen ($H_2$) is reacted with nitrogen ($N_2$):

$$H_2 \quad + \quad N_2 \quad \longrightarrow \quad NH_3$$

hydrogen + nitrogen ⟶ ammonia

Can you see that the numbers of atoms that react do not balance the number of atoms that are in the product?

## Balancing equations

Count the atoms in the reactants and the products of the equation for the formation of ammonia.

There are two atoms of hydrogen and two atoms of nitrogen producing three atoms of hydrogen and only one atom of nitrogen. This is impossible!

The solution is to increase the number of molecules of hydrogen being used and also increase the number of molecules of ammonia being produced.

There have to be two ammonia molecules made, because two nitrogen atoms are being used. Two ammonia molecules would contain six hydrogen atoms. That means three hydrogen molecules are needed. This is what the balanced equation looks like:

$$3\,H_2 \quad + \quad N_2 \quad \longrightarrow \quad 2\,NH_3$$

## Balancing symbol equations

Writing a number in front of a chemical formula tells chemists how many molecules take part in a reaction. This is crucial for making ammonia. It tells chemists they need three times as much hydrogen as nitrogen for the reaction to work properly. If there is not enough hydrogen, some of the nitrogen will be wasted. This would waste money, making ammonia more expensive.

The next example shows how ammonium sulfate (a fertiliser) is made by adding sulfuric acid to ammonia:

$$NH_3 \quad + \quad H_2SO_4 \quad \longrightarrow \quad (NH_4)_2SO_4$$

In the reactants: one nitrogen atom, five hydrogen atoms, one sulfur atom, four oxygen atoms.

In the products: two nitrogen atoms, eight hydrogen atoms, one sulfur atom, four oxygen atoms.

So the reactants need one more nitrogen atom and three more hydrogen atoms.

Therefore, there must be two ammonia molecules, not one.

So the balanced equation is:

$$2\,NH_3 \quad + \quad H_2SO_4 \quad \longrightarrow \quad (NH_4)_2SO_4$$

| Symbol | Atoms in reactants | Atoms in products |
|--------|--------------------|--------------------|
| N | 1 | 2 |
| H | 5 | 8 |
| S | 1 | 1 |
| O | 4 | 4 |

## Summary questions

1 a What reactants are used to make ammonia?
  b Name a fertiliser can be made from ammonia.

2 Hydrogen gas ($H_2$) is produced by reacting methane ($CH_4$) with steam ($H_2O$). Carbon dioxide ($CO_2$) is also produced.
  a Write a word equation to show this.
  b Write a balanced symbol equation to show this.  [H]

3 Balance the following equation, showing the reaction between hydrochloric acid and sodium carbonate:
  $$HCl + Na_2CO_3 \longrightarrow H_2O + CO_2 + NaCl \quad [H]$$

### Key points

- Chemical equations are used to show how chemicals react and what is produced.

- Balanced chemical equations help chemists calculate how much raw material they need. [H]

# 2.10

# The cost of a product

## Learning objectives

- What factors affect the cost of producing chemicals?
- How does the cost of raw materials affect us all?
- How can we calculate the cost of a product, and the cost of any waste?

When making new products, manufacturers need to consider the amount of money being spent on raw materials. As we have seen, any waste will cost the manufacture money. Other factors also affect the cost of making new products. They include paying workers, using energy and building the facilities to make the product safely.

### Value for money

Reducing waste and the costs of raw materials affect consumers. For instance, if ammonium nitrate can be made cheaply, it can be sold for less. This means it is cheaper for farmers to fertilise their land. In turn, that means more crops can be grown for less money, keeping the price of food down. In this way, industrial chemistry affects us all as consumers.

As well as affecting the price of fuel, the cost of separating hydrocarbons from crude oil will affect a huge number of products. Everything from plastics to makeup uses hydrocarbons as raw materials. This means that the cost of fractional distillation of crude oil will be passed on to the consumer on a wide range of products.

### Factors affecting the cost of a product

There are a few factors to think about when producing materials: the cost of the reactants, the price the products can be sold for, the cost of any waste, and 'overhead' costs. Overhead costs include the energy needed in production, maintaining a safe plant, any taxes, the wages of workers and the cost of transporting the final product.

So, the financial cost of making a product = cost of reactants + cost of waste + overhead costs.

To make a profit, these costs must be less than the price the final product is sold for:

Profit = price the product is sold for − total costs

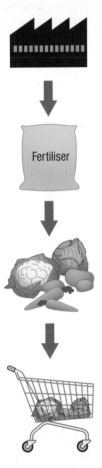

Figure 1 The cost of producing fertiliser affects the cost of our food in the supermarket

### Maths skills

In the following example, a factory is making chemical X from chemicals A and B:

- X can be sold for £1500 per tonne.
- A costs £800 per tonne; B costs £200 per tonne.
- Overheads cost £5000 per day.
- The factory makes 20 tonnes of X every day.
- To make X, the factory uses 15 tonnes of A and 5 tonnes of B every day.

The total daily cost of A would be 15 × £800 = £12 000

The total daily cost of B would be 5 × £200 = £1000

With no waste, the total daily cost would be:

£12 000 A + £1000 B + £5000 overheads = £18 000

20 tonnes of chemical X is worth 20 × £1500 = £30 000

So in this example, the factory makes a profit of £30 000 − £18 000 = £12 000 every day.

## Activity

### Calculating the cost

A chemical plant makes 80 tonnes of ammonium nitrate every day. The ammonium nitrate can be sold for £3000 per tonne.

The equation for the reaction is:

ammonia   +   nitric acid   $\longrightarrow$   ammonium nitrate

$NH_3$   +   $HNO_3$   $\longrightarrow$   $NH_4NO_3$

### The costs of the product

**a** If the factory used 17 tonnes of ammonia every day, how much nitric acid would it use every day?

**b** If ammonia costs £4000 per tonne, how much will the ammonia cost each day?

**c** If nitric acid costs £1200 per tonne, how much will the nitric acid cost each day?

### Running costs

The energy needed, worker wages, maintenance and other costs of running the chemical plant are £16 400 per day.

**d** What is the total cost of running the chemical plant each day?

**e** How much money would be wasted in a week if 1 tonne too much ammonia was used every day?

### Is it making money?

**f** How much money can be made in one day from selling the ammonium nitrate?

**g** Is the factory making a profit? If so, how much per day?

**Figure 2** This UK chemical plant makes chemicals for use in new products. In order to make the chemicals as cheaply and as safely as possible, chemists must work out the correct amounts of materials to use.

## Summary questions

1 Copy and complete using these words:

*cost   transportation   cheaper   waste   more   overheads*

The cost of making a product depends on the ............. of the reactants, ............., and the cost of any ............. materials. The less waste, the ............. the product is to make. Overheads include wages for workers, the cost of energy and other costs such as taxes or ............. To make a profit, the product must be worth ............. than the total cost of making it.

2 List the costs of running a chemical plant.

3 Name **two** products that would be cheaper if the cost of fractional distillation could be reduced.

4 How can the cost of ammonium sulfate affect the price of a bag of crisps?

### Key points

- Using the right amount of raw materials prevents wastage and saves money.

- The cost of raw materials affects a wide range of consumer products.

- The cost of industrially made products depends on the costs of the reactants, overheads and waste.

# Summary questions

**1** Match each word to its definition.

| | |
|---|---|
| Atom | A charged particle |
| Compound | Made of more than one type of atom not chemically combined |
| Element | The building block of all chemicals |
| Ion | Made of one type of atom |
| Mixture | Made of two or more atoms stuck together |
| Molecule | Made of two or more types of atom chemically combined |

**2** The mass number of carbon is 12. Its atomic number is 6.

**a** How many protons does a carbon atom have?

**b** How many neutrons does a carbon atom have?

**c** How many electrons does a carbon atom have?

**3** Particles can be either molecules, ions or atoms.

**a** What is the difference between an atom and an ion?

**b** What are the two types of ion?

**c** Decide whether the following particles are atoms, ions or molecules.

$Cu$   $Ca^{2+}$   $H_2$   $Fe$   $N_2$   $Cl^-$   $Na^+$   $H_2O$

**4** We can use some materials from the Earth straight from the ground. Others must be separated from other materials first.

**a** Put the following materials into two groups, straight from the ground or needing separation:

salt   gold   limestone   petrol   sulfur   marble

**b** Crude oil can be separated. This is because the different materials in it have different boiling points. Why can rock salt be separated into rock and salt?

**5** Metals are obtained from rocks called ores.

**a** What element often has to be removed from ores in order to extract the pure metal?

**b** What waste gases are produced when iron and lead are produced?

**c** Why is carbon used as a reducing agent for lead and iron oxides?

**6** **a** Describe some important products made using gases extracted from the atmosphere.

**b** Which gases can be extracted from the atmosphere and used to make other materials?

**7** The materials we extract from the Earth have impacts on the environment, people and the economy.

**a** Who would be affected by the opening of a new limestone quarry? What are the positive and negative impacts?

**b** How would the environment be affected by the opening of a new quarry?

**8** When calcium carbonate ($CaCO_3$) is heated, it decomposes into calcium oxide ($CaO$) and carbon dioxide ($CO_2$).

**a** Write a word equation to show this happening.

**b** If 88 g of calcium carbonate releases 36 g of carbon dioxide when it decomposes, how much calcium oxide is left?

**c** Use a diagram to explain how mass is conserved during this reaction. [H]

**9** Explain the effect the following factors would have on the cost of making a product:

**a** more automatic machinery

**b** reducing energy costs

**c** using too much of one reactant

# AQA Examination-style questions

**1** The diagrams show the atoms in various substances.

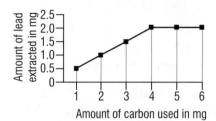

A    B    C    D    E

**a** Explain which diagram represents a mixture of different compounds. *(3)*

**b** Explain which diagram represents molecules of one element. *(3)*

**2** Crude oil is a mixture of hydrocarbons that need to be separated before use.

**a** Describe the process used to separate crude oil. *(4)*

**b** Name **two** liquids separated from crude oil that can be used in a car. *(2)*

**3** Use chemicals from the box to answer the questions.

> carbon   gold   helium   iron
> nitrogen   sulfur   water

**a** Name and explain which chemical is a compound. *(3)*

**b** Name an element that can be used straight from the ground and explain why it does not need to be processed first. *(3)*

**c** Give the chemical symbol for the element that needs to be extracted from its ore. *(1)*

**4** The table shows the price of some metals that are used to make brass.

| Metal | Price in £ per kg |
|-------|------------------|
| Copper | 0.96 |
| Iron | 0.14 |
| Tin | 2.28 |
| Zinc | 0.25 |

**a** The brass used for making metal hoses is called low brass. It is 80% copper and 20% zinc. Calculate the cost of making 10 kg of low brass.

£1.21   £7.68   £8.18   £12.24 *(1)*

**b** Aich's brass is used for boats as it is corrosion resistant. It is made from 60.66% copper, 36.58% zinc, 1.02% tin and the rest is iron. Calculate how much iron is in Aich's brass. *(1)*

**5** Scientists have studied the structure of the atom so they can understand how reactions occur.

**a** Draw and label a lithium atom: $^{7}_{3}Li$ *(4)*

**b** Describe how a lithium atom is turned into a lithium ion: $Li^{+}$. *(1)*

**6** A manufacturing company was given many small samples of lead oxide.

They did an experiment to find how much carbon to use to extract the lead from each sample.

Their results are shown in the graph below.

**a** How much lead was in the samples? *(1)*

**b** Suggest **two** things that the company could have done to make sure they collected accurate results. *(2)*

**c** Suggest and explain the amount of carbon the company should use. *(2)*

**d** *In this question you will be assessed on using good English, organising information clearly and using specialist terms where appropriate.*

Describe the reaction that the company could have used to separate lead from its ore. *(6)*

**7** Air is a mixture of gases that can be separated. First carbon dioxide, water and dust are removed. The air is then compressed, cooled and allowed to expand.

This is repeated until it turns into a mixture of liquids all with different boiling points.

| Gas | Boiling point in °C |
|-----|---------------------|
| Argon | −185 |
| Nitrogen | −195 |
| Oxygen | −182 |

Use information from the table to explain how pure oxygen can be separated from liquid air. *(3)*

**[H]**

# Life on our planet

In Unit 1, Theme 2 you will work in the following contexts, covered in Chapters 3 and 4:

## Life on our planet

### How can plants and animals survive in extreme environments?

Plants and animals have specific characteristics that allow them to live in a particular environment. These are known as adaptations. For example, to survive in the extreme cold of the Arctic most mammals are covered in layers of fat and blubber, and have very thick fur. Plants and animals that live in desert environments have different characteristics. Here, animals are adapted to transfer energy in order to cool down – for example, by not having much fur.

### Do we look like our ancestors?

The further back in time you go, the more different we look from our ancestors. This is true for all species. The wide variety of life present on the Earth has evolved over time and is still changing. Organisms are continually evolving to become better adapted to their environment. This process is known as natural selection.

## Biomass and energy flow through the biosphere

### How do organisms gain energy?

Plants obtain their energy from the Sun in the form of light. This is transferred into chemical energy during photosynthesis. Animals can only gain energy by eating other organisms. Some of the energy is used to power body reactions, such as movement. Some is used for growth, which increases an organism's biomass.

## How do scientists monitor the flow of energy and biomass through the biosphere?

Ecologists study the movement of energy and biomass within food chains. To achieve this, data is collected on the numbers and sizes of organisms within a feeding relationship. Generally, as you move along a food chain the size of organism increases. However, fewer and fewer organisms exist at each trophic level.

## The importance of carbon

### Why is carbon so important?

Carbon is an essential element in all organisms. In fact, it is the major element within your body. Carbon forms the basis of all organic molecules, such as carbohydrates, fats and proteins. These are the essential building blocks of life.

Carbon is constantly cycled through the environment. For example, it is removed from the atmosphere when plants photosynthesise, and returned when organisms respire.

### How are humans affecting the carbon cycle?

Carbon is stored underground, in fossil fuels. It is also present in the Earth's atmosphere, in the form of carbon dioxide which is also dissolved in the oceans.

Over the past 40 years, large areas of forest have been cleared to provide extra space for farming. This means that there are now fewer trees to remove carbon dioxide from the atmosphere. In addition, many countries burn fossil fuels to generate electricity and power transport. These processes release extra carbon dioxide into the atmosphere. Both of these processes are disrupting the natural carbon cycle.

Bacteria and fungi play an essential role in decomposition. Decomposers are nature's recyclers and form an essential part of the world's ecosystems. The nutrients they release are used by plants for growth. Without decomposers, dead remains would not be broken down. In fact, without them the world would still be covered with the dead remains of dinosaurs!

In order to survive, species need to be able to cope with changes to the environment. Through mutations, individuals in a species can develop characteristics that provide survival advantages. For example, some bacteria have developed a resistance to antibiotic drugs. This is an advantage for the bacteria, as they cannot be killed using traditional treatments. These bacteria are therefore more likely to survive, and pass on this advantage to the next generation. Over recent years, increased numbers of resistant bacteria have been detected. This is an example of natural selection going on today.

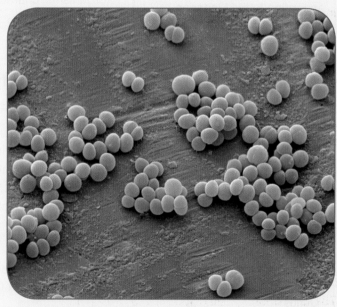

Agricultural scientists study ways to maximise plant growth. For example, they suggest the ideal conditions in which a crop should be grown. This results in higher crop yields and cropping over longer periods of the year, helping keep supplies high and costs low. Auxins are plant hormones responsible for controlling plant growth. For example, they make plants grow towards the light (phototropism). Farmers use products containing auxins to control weed growth, and to help ripen fruit.

# 3.1

# Classification ⓚ

## Learning objectives

- What is classification?
- Why are organisms classified?
- How can we classify organisms

## ?? Did you know ... ?

The diagram below shows how humans are classified according to the Linnaean classification system. This is how we get our name *Homo sapiens*. A species name is specific. It refers to a single organism rather than a group.

| | |
|---|---|
| Kingdom – *Animalia* | Broadest category |
| Phylum – *Chordata* | |
| Class – *Mammalia* | |
| Order – *Primates* | |
| Family – *Hominoidea* | |
| Genus – *Homo* | |
| Species – *sapiens* | Most specific category – contains only one type of organism |

**Classification** means sorting things into groups based on similar features. Lots of everyday objects are classified. For example, in a supermarket, semi-skimmed milk is classified as a dairy product. If you wanted to find a similar product such as cream, you know it would be located nearby.

**a** What does 'classify' mean?

Living organisms are classified into **taxonomic groups**. All species within a taxonomic group share similar characteristics. This system of classification was introduced by Carl Linnaeus in the eighteenth century. It is now used by scientists around the world.

There are seven main groups. They are arranged in order, from **kingdom** – the broadest category (organisms share some characteristics), to **species** (organisms' characteristics are almost identical).

Within each kingdom, organisms are further subdivided into smaller and smaller groups.

## Ways to classify living things

The main groups of living things include:

- Plants – organisms that make their own food by photosynthesis. For example, flowering plants.
- Animals – organisms that cannot make their own food. For example, insects.
- Microbes – which would include fungi and single-celled organisms.

**b** What is the main characteristic of the plant kingdom?

Organisms in the animal kingdom can be divided into two groups depending on whether or not they have a backbone:

**1 Vertebrates** – animals with a backbone.
**2 Invertebrates** – animals without a backbone.

**c** What is the difference between a vertebrate and an invertebrate?

The table shows the characteristics of the five vertebrate groups.

| Characteristics | Vertebrates | | | | |
|---|---|---|---|---|---|
| | **Mammals** | **Birds** | **Fish** | **Reptiles** | **Amphibians** |
| **Blood** | Warm blooded | Warm blooded | Cold blooded | Cold blooded | Cold blooded |
| **Reproduction** | Live young | Lay hard-shelled eggs | Lay eggs in water | Lay soft-shelled eggs | Lay jelly-coated eggs in water |
| **Skin covering** | Have hair | Have feathers | Have wet scales | Have dry scales | Have moist skin |

## Why classify organisms?

Scientists classify organisms for a number of reasons:

1 To name and identify species – it makes it easier to find out which species an organism belongs to if everything is organised.
2 To predict characteristics – if several members in a group have a particular characteristic, another species in the group may have the characteristic.
3 To find evolutionary links – species in the same group probably share characteristics because they have evolved from a common ancestor.

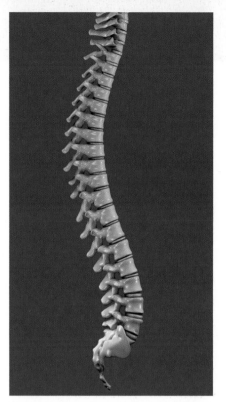

**Figure 1** The name vertebrate comes from the vertebrae present in the spinal column

### Activity

#### The Natural History Museum

The Natural History Museum in London, houses the **largest** and most **important natural history collection** in the world. This diverse collection has been gathered over the last 400 years. It contains over **70 million** specimens ranging from microscopic specimens to dinosaur skeletons. The specimens are organised into 'collections' – groups of items that

have something in common. These are constantly being reorganised and developed. They try to take into account the latest scientific thinking on the classification and relationships between organisms.

The collections are used by scientists to investigate the natural world. They also provide a point of reference and authority for wider investigations by scientists around the world. By studying the specimens, scientists gain knowledge of animals and plants. They can also discover the processes that have shaped the world and our solar system.

● Find out about the classification work carried out at the Natural History Museum.
● Write a leaflet for fellow students explaining the work scientists carry out there.

**?? Did you know … ?**

It has been estimated that approximately two-thirds of all living species are insects. Only around 1 per cent of all animal species are larger than a bumble bee!

**∞ links**

*Variation exists between members of the same species but their main characteristics are the same. For more information on variation within species see 6.2 Variation.*

### Summary questions

1 Copy and complete using these words:

*groups   organisms   animals   characteristics
plants   classification   taxonomic*

............. means sorting things into ............. .

............. are sorted into ............. groups based on similar physical ............. .

The main groups or kingdoms include ............. and ............. .

2 How could you tell the difference between an amphibian and a reptile?

3 Why is it important that scientists around the world use the same classification system?

### Key points

● Classification sorts things that share similar features into groups.

● Organisms are classified into groups to aid naming and identification.

● Organisms are classified into groups based on their physical characteristics.

### 3.2

# Using evolutionary and ecological relationships

## Learning objectives

- What is an evolutionary tree?
- How do predator and prey populations depend on each other?

### ∞ links

*For more information about evolution and fossils go to 3.5 Evolution.*

By studying the characteristics of organisms, scientists know how species are related to each other. The same classification system is used all over the world so that scientists can share their research. This means that links between different organisms can be seen, even if they live on different continents.

## Evolutionary relationships

Fossil records have enabled scientists to produce **evolutionary trees**. These are branched diagrams that show how different species have evolved from a common ancestor. Evolutionary trees are produced by looking at similarities and differences in a species' physical characteristics and genetic makeup.

**a** What does an evolutionary tree show?

The elephant's evolutionary tree shows that all species of elephant have evolved from a palaeomastodon. This organism was alive at the time of the dinosaurs. Over time, the appearance of an elephant has changed. That's because it has adapted to the environment it now lives in. Today, only two species of elephant survive – the Indian elephant and the African elephant.

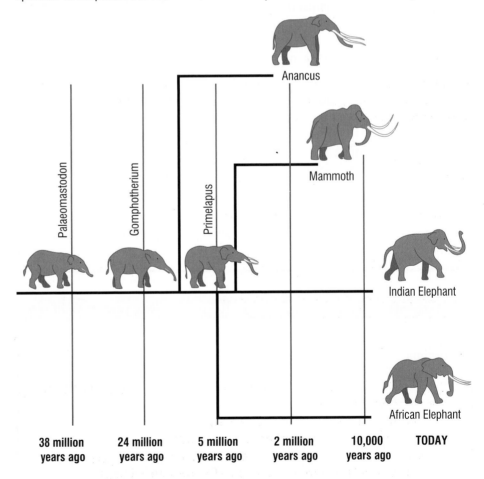

**Figure 1** The elephant's evolutionary tree

### AQA Examiner's tip

To answer questions about evolutionary trees, you need to look both at the drawings of the organism, and the timeline. Often, the organisms that are alive today are drawn at the top of the diagram. The common ancestor from which all the species develop is found at the bottom. This will be where all the branches meet. If the species is not present in the highest level of the diagram then it is now extinct.

**b** Around what time did Anancus become extinct?

**Figure 2** The hare is the lynx's preferred food source

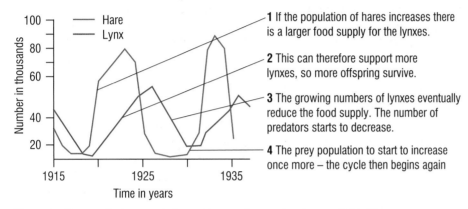

**1** If the population of hares increases there is a larger food supply for the lynxes.

**2** This can therefore support more lynxes, so more offspring survive.

**3** The growing numbers of lynxes eventually reduce the food supply. The number of predators starts to decrease.

**4** The prey population to start to increase once more – the cycle then begins again

**Figure 3** Changes in the populations of lynx and snowshoe hares, 1915–35

## Ecological relationships

Scientists also study how species depend on each other. These interactions are known as ecological relationships. One example is the relationship between a predator and its prey.

**Predator** species need to be adapted to hunt and kill, to ensure they catch enough food to survive. **Prey** species must be adapted to escape their predators. These features ensure enough organisms survive for the species to continue.

**c** What is the difference between a predator and its prey?

### Summary questions

**1** Copy and complete using these words:

*ancestor   evolutionary   evolved*

Scientists use ............ trees to show how organisms have ............ from a common ............ .

2 Using the elephant's evolutionary tree on the previous page, answer the following questions:

**a** What is the elephant's common ancestor?

**b** Which two species of elephant are alive today?

**c** Which species of elephant was alive around 24 million years ago?

**d** What was the last species of elephant to become extinct?

**3** Describe the relationship between aphid (prey) and ladybird (predator) populations.

### Key points

- An evolutionary tree shows the evolutionary relationship between species. It shows they have evolved from a common ancestor.

- Predator populations depend upon the population of their prey. If the number of prey organisms decreases, so does the number of predator organisms.

## 3.3

# Competition

### Learning objectives

- What is a habitat?
- What do animals need to survive?
- What do plants need to survive?

**Figure 1** Limpets have to cope with regular changes in sea level. When they are underwater, limpets move over rocks eating small pieces of seaweed. When the tide goes out, they cling tightly to the rock. This stops them being washed away or being eaten by predators.

A **habitat** is the place where an organism lives. There are many different habitats including ocean, desert, forest, pond or even a garden.

**a** Name **three** examples of a habitat.

Each habitat has different **environmental conditions**. These include temperature and amount of rainfall. The environmental conditions in most habitats vary throughout the day and throughout the year:

- **Daily changes** – these include changes in light levels and temperature. Night time is darker, and generally colder, than day time. To cope with this change most animals sleep during the night. However, nocturnal animals such as foxes and owls use this as an advantage and hunt during the night.
- **Seasonal changes** – these include changes in temperature and rainfall. For example, in the winter it is colder and the days are shorter. There is often more rainfall. Some animals, like hedgehogs, cope with this change by hibernating.

**b** Why do most animals sleep during the night?

An **ecosystem** is the name given to a habitat and all the living organisms that live there. To survive, the plants and animals need a number of different materials from their surroundings. If materials are limited, organisms have to **compete** for these resources. For example, plants compete for access to light. In many cases, only the strongest species will survive.

**c** Why do plants and animals compete for resources?

### What do plants need to survive?

1 Sunlight ⎫
2 Water ⎬ – Needed for **photosynthesis**, to produce food for growth
3 Nutrients – Needed for healthy growth

### What do animals need to survive?

1 Food ⎫
2 Water ⎬ – To grow
3 Mates – To reproduce
4 Suitable territory – For safety and shelter

### A pond ecosystem

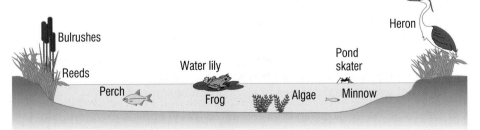

**Figure 2** The animals and plants found in a pond ecosystem

Many kinds of plant and animal life can be found in a pond. The different **populations** of species live together in a **community**. The variety and number of organisms in this community are determined by the amount of oxygen, light and shelter that is available. Some plants and animals can live in the pond, others live on or near the pond.

> **d** Name a microorganism, plant and animal species that can live in a pond habitat.

The community also depends on the way plants and animals live together in the pond. Organisms may compete for resources such as light. If the water lily covers part of the surface of the pond, other plants living below it will die. That's because not enough light can reach the plants below.

Other organisms in a community may provide a food source. For example, the heron may feed on the minnow population. This means that the perch may no longer be able survive in this habitat as its food source has been used up.

Ecosystems are in a constant state of balance. Small changes can have dramatic effects.

## Practical

### Surveying organisms using a line transect

A line transect is a sampling technique that compares the conditions in a habitat with the species present at that point. A number of samples are taken along a pre-marked path. The range of data collected allows scientists to link environmental factors with the type and number of a species.

Carry out a line transect of a habitat – for example, you could look at an area of your school grounds. Stretch a measuring tape for a distance of 10 metres. At every metre along the tape:

1 List the plants and animals that are present at each metre along the line.
2 Note the environmental conditions at each sampling point. For example, the light level, or how trampled the ground is.
3 How do the environmental conditions affect the distribution of plants and animals along the line?

## Summary questions

1 Match the following terms to their definition:

ecosystem      place where an organism lives
population      the conditions that surround an organism
environment    the habitat and living organisms within it
habitat        the different species living in a habitat
community       numbers of the same species living in a habitat

2 **a** What do plants need to survive?
  **b** What do animals need to survive?

3 Most fish that live in a river cannot survive in an ocean habitat. What is the main difference between these two habitats?

4 **a** What are the differences in habitat between the surface and the bottom of a pond?
  **b** How might these differences affect the types of organisms that live there?

### Key points

● A habitat is a place where a living organism lives.

● Plants need sunlight, water and nutrients to survive.

● Animals need food, water, mates and a suitable territory to survive.

# 3.4 Adaptations

## Learning objectives

- What is an adaptation?
- How are animals and plants adapted for life in the desert?
- How are animals and plants adapted for life in the Arctic?

**Figure 1** A cactus has many ways of preventing water loss to allow it to survive in a desert

**Figure 2** A camel has many adaptations that allow it to survive in a hot environment

All organisms have characteristics that allow them to live successfully in their habitat. These characteristics are known as **adaptations**.

**a** What is an adaptation?

## Life in the desert

Deserts receive very little rainfall, making them very dry. They are usually hot during the day and extremely cold at night. This makes a desert a very harsh habitat to live in. Cacti and camels have many adaptations that help them to survive there.

**b** Describe the habitat of an organism that lives in the desert.

## Cactus adaptations

- thick waxy coating – reduces water loss through evaporation
- succulent plants – stores water in the stem
- covered in spines (modified leaves) – stops predators eating them
- stem carries out most of the photosynthesis – reduces the need for leaves
- widespread root systems – collects water from a large area.

**c** How does a cactus minimise water loss?

## Camel adaptations

- large flat feet – stops them sinking into the sand
- do not sweat or urinate very often – reduces water loss from the body
- long eyelashes and slits for nostrils – prevents sand from entering the body
- large surface area to volume ratio – maximises heat loss
- fur – thick on top of the body for shade, thin everywhere else to maximise heat loss.

### Practical

**Measuring surface area to volume ratios**

The size and surface area of many animals are adapted to suit the environment in which they live. For example, camels have a large surface area to volume ratio to cause maximum heat loss.

Place $50\,\text{cm}^3$ of water at $70\,°\text{C}$ into beakers of different surface areas (sizes). Take the temperature of the water every 30 seconds for 5 minutes.

Is there a relationship between the surface area of the flask and the rate of heat loss?

## Life in the Arctic

The Arctic is very cold – much of the land is covered in snow or ice all year. The winter is extremely cold and long, and the summer is short and cool. It is often very windy and there is little rainfall. Polar bears and prairie crocuses are examples of organisms that have adapted to live in these harsh conditions.

### Prairie crocus adaptations

- plants are small and near to the ground
- plants grow close together} helps protect plants from cold and strong winds
- stems, leaves and buds covered in fine hairs – provides insulation
- produce flowers quickly in short growing season – maximises chance of reproduction
- small leaves – helps reduce water loss.

  **d** How do prairie crocuses prevent water loss?

### Polar bear adaptations

- white fur – camouflages bear from its prey
- thick layers of fat and fur – provide insulation
- large feet – stops them sinking into snow
- hairs on sole of feet – prevents slipping and provides insulation
- small surface area to volume ratio – reduces energy loss
- sharp claws and teeth – weapons to catch and eat prey.

**Figure 4** A polar bear has many adaptations that allow it to survive in a cold environment

  **e** What adaptations does a polar bear have to keep it warm?

### Living in extreme environments

Microorganisms have been found living in very cold places, such as the Arctic, and very dry environments, such as the desert. This may not be that unusual. However, **extremophiles** are microorganisms that can survive in extreme environments where no other kind of organism could exist. These include volcanic vents, found deep under the oceans, and highly acidic hot springs!

**Figure 3** A prairie crocus has adapted to survive cold temperatures and strong winds

### AQA Examiner's tip

Think about the habitat that an organism lives in so you can explain its adaptations.

### ??? Did you know ...?

The snowshoe hare changes the colour of its fur with the seasons. In the summer it has brown fur to camouflage with soil and tree trunks. In winter it has white fur to camouflage with snow. This adaptation helps it remain hidden from predators.

### Key points

- An adaptation is a specific characteristic that allows an organism to live in a particular habitat.
- Plants and animals that live in the desert have special adaptations to enable them to cope with the heat and lack of water.
- Plants and animals that live in the Arctic have special adaptations to help them cope with the cold, windy conditions.

---

### Summary questions

1  Copy and complete using these words:

   *habitat  camouflaged  prey  adaptations*
   *predators  characteristics*

   ............ are specific ............ that allow organisms to live in a particular ............ .

   Many animals are ............ so that they blend in with their environment. This helps keep them hidden from ............, or allows them to sneak up on their ............ .

2  What is the relationship between surface area and rate of cooling?

3  How is a seal adapted to living in the ocean?

# 3.5  Evolution

## Learning objectives

- What is evolution?
- What is natural selection?
- What is a fossil?

 **Did you know ... ?**

You share about 95% of your genes with a gorilla and 50% with a banana! This is because all living things evolved from the same ancestor millions of years ago.

**Figure 1** Charles Darwin

Scientists believe that all living organisms have gradually developed, from a common ancestor over millions of years. This process is called **evolution**.

## Darwin's theory of evolution

Charles Darwin was a famous English scientist. In 1859 he published the theory of evolution in his book on the origin of species. He based it on his observations of variation in plants and animals on a voyage around the world. Darwin's ideas caused controversy as they conflicted with religious views on creation.

Charles Darwin's theory of evolution states that all species have evolved from simple life forms. These simple organisms lived in water more than three billion years ago. They were similar to bacteria found today. These organisms evolved to become more complex. Eventually, organisms developed that could live on land and in the air.

## Evidence for evolution – fossils

**Fossils** were formed when animal and plant remains were preserved in mud. Over millions of years the mud turned into rocks. The fossil record provides most of the **evidence** for evolution. Different fossils show that organisms have gradually changed over time to become more adapted to their environment. The best adapted organisms survived to reproduce. This process is known as **natural selection**. However, the fossil record is not complete as not all organisms fossilise well. Many fossils have also been destroyed by the Earth's movements, or lie undiscovered. This means that some evidence for evolution has not yet been discovered.

**a** What is a fossil?

## Natural selection

Organisms evolve through the process of natural selection. Natural selection takes years (sometimes millions of years) to occur, as it has to take place over a number of generations. It follows the steps below:

Organisms in a species show a wide range of variation
(caused by genetic differences).

↓

The organisms with the characteristics that are most suited
to the environment are most likely to survive and reproduce.
This is often referred to as 'survival of the fittest'.

↓

Genes from successful organisms are passed
to the offspring in the next generation.

AQA **Examiner's tip**

Make sure you know the meanings of the terms 'natural selection' and 'evolution'.

## links

*Farmers exploit the process of natural selection when they selectively breed their plants or animals. You can find out more in 13.1 Selective breeding.*

This process is then repeated many times. Living organisms are continually evolving to become better adapted to their environment. This means many species are still evolving by natural selection.

**b** What does 'survival of the fittest' mean?

## Antibiotic resistance

Antibiotic-resistant bacteria cause many problems in hospitals. MRSA is an example species.

Bacteria reproduce very rapidly and so can evolve in a relatively short time. When bacteria divide, their DNA can be damaged or altered. This usually results in the bacteria dying. However, the **mutation** (altered DNA) is sometimes an advantage for the bacteria. It can increase their chance of survival. For example, the mutation may cause resistance to an antibiotic. The antibiotic will no longer be able to destroy the bacteria.

**c** What term is used to describe DNA that has been damaged or altered?

## Peppered moths

Originally, most peppered moths in Britain were pale coloured. This made them camouflaged against tree trunks. A mutant form of these moths is dark coloured. These were easily seen by birds and eaten. The pale moths were therefore more likely to survive and reproduce.

In industrialised areas soot started to coat trees, turning the bark black. This meant that the black moths were now better camouflaged, and so more of them survived. After several years, dark peppered moths became more common in towns and cities than pale moths.

## Extinction

The fossil record shows that many species have become **extinct** since life began. If a species is poorly adapted to its environment it will not survive and will eventually become extinct. Extinction occurs naturally but is often made more likely through human activity. Factors that can cause a species to become extinct include:

- changes to environmental conditions
- new diseases
- new predators and competitors
- destruction of habitat
- human use of organisms for food or materials.

### Summary questions

1 Copy and complete using these words:

*selection   fossils   evolved   adapted   evidence*

Organisms have ............ from a common ancestor. ............ for this is gained from ............ . These show that organisms have gradually become better ............ to their environment, through the process of natural ............ .

2 Why might a species become extinct?

3 Which type of peppered moth would you be most likely to find near a steel factory? Explain your answer.

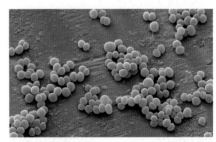

**Figure 2** Methicillin-resistant *Staphylococcus aureus* (MRSA) – a hospital superbug

### ∞ links

*For more information about bacteria, the use of antibiotics and antibiotic resistant strains see 10.2 Antibiotics*

### Key points

- Evolution is the theory that all organisms have gradually evolved from a common ancestor over millions of years.

- Natural selection occurs because only the organisms with the characteristics most suited to their environment survive and reproduce. They pass on these characteristics to their offspring through genes.

- A fossil is formed when a plant or animal's remains are preserved in rock. The fossil record provides evidence for evolution.

## 3.6 Plant growth

### Learning objectives

- What environmental factors affect plant growth?
- How do plants respond to their environment?
- What is the difference between phototropism and gravitropism?

### AQA Examiner's tip

Remember that auxins can speed up the growth of a plant or slow it down. Generally, auxins slow down the growth of roots but speed up the growth in other parts of the plant.

Plant growth is affected by a number of environmental (external) factors. These include light, temperature, day length and gravity.

Plants respond to their environment using **auxins**. These chemicals are plant hormones.

Plants detect stimuli in their environment, and can respond by growth – **tropism**. If a part of a plant grows towards a stimulus it is called a positive tropism. If it grows away from the stimulus it is called a negative tropism.

**a** A plant grows towards water. Is this a positive or negative tropism?

### Light stimulus

**Phototropism** means growing in response to light. When a stem grows towards light, the plant can photosynthesise more. This means more food is produced for the plant, so it can grow faster. This increases the plant's chances of survival.

**b** Why is it beneficial for plants to grow towards a light source?

### Gravity stimulus

**Gravitropism** means growing in response to gravity. It is important for a plant's roots to grow downwards, as growing deeper into the soil helps to provide anchorage. It normally also takes the roots nearer to a source of water.

---

### Practical

#### Investigating gravitropism

Place two sets of cress seeds on dampened cotton wool in agar plates. Once the seeds have germinated, place one of the plates in a clinostat; this device rotates the seeds to cancel the effect of gravity. Leave the second plate adjacent to the clinostat – this is the control.

After a few days, examine the growth on the two plates.

- Why are they different?

---

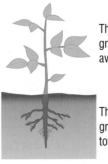

The shoot is negatively gravitropic – it grows away from gravity

The root is positively gravitropic – it grows towards gravity

**Figure 1** Plants grow in response to gravity

## How do auxins affect plant growth?

Auxins are made in cells near the tips of plant shoots or roots. They make these parts of the plant grow faster than others. Auxins are used by plants to enable them to grow towards, or away from a stimulus.

**c** How do auxins affect plant cells?

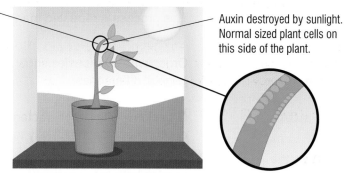

More auxin present on shaded side of plant. This causes cells to lengthen on this side of the plant.

Auxin destroyed by sunlight. Normal sized plant cells on this side of the plant.

**Figure 2** Auxins cause plant shoots to grow towards the Sun

## The growing season

All plants have a growing season. This is when conditions are suitable for their growth. These conditions include temperature and the light intensity that a plant receives. Both of these factors are affected by day length. Most plants cannot grow at temperatures below 6 °C unless they are specially adapted.

Farmers and commercial growers carefully study the growing season of a plant. By creating the ideal conditions for growth in a greenhouse, crops can be grown quickly all year round. Techniques that are used include:

- artificial lighting – to allow photosynthesis to continue after daylight hours
- high carbon dioxide levels – to increase the rate of photosynthesis
- heating – to ensure photosynthesis occurs at a reasonable rate.

Paraffin lamps are often used inside greenhouses as they give off carbon dioxide as well as light and heat.

**d** Name two techniques farmers can use to increase plant growth.

### Summary questions

1 Copy and complete using these words:

*gravitropism environmental temperature light phototropism stimulus auxins*

Many ............ factors have an effect on plant growth. These include ............ and ............ intensity.

Plants react to a ............ in their environment using ............ . The word ............ means growing towards gravity and ............ means growing towards light.

2 Name a stimulus to which a plant shows:
   **a** positive tropism
   **b** negative tropism.

3 If you cut the tips off of a plant shoot, why does the plant not grow towards the light?

### ??? Did you know ...?

Farmers also use plant hormones to help crops to grow effectively. These include:

- Killing weeds – some weedkillers contain growth hormones. They selectively kill weeds by making them grow too fast.
- Promoting root growth – rooting powder (a growth hormone) is pasted onto plant cuttings to promote root growth.
- Ripening fruit – ethene (a plant hormone) is sprayed on fruits so they ripen quicker.
- Producing seedless fruit – hormones can be used to produce seedless fruit, which many people prefer to eat.

### Key points

- Light and temperature (day length) and gravity are some environmental factors that affect plant growth.

- Plants respond to their environment using auxins – plant hormones.

- Phototropism means growing towards the light. Gravitropism means growing towards gravity.

# Summary questions

**1** To aid in the research and understanding of organisms, scientists classify organisms.

    **a** What is meant by the term *classification*?

    **b** Using the kingdoms (groups) in the table below, classify the following organisms:

        *oak    horse    jellyfish    yeast*
        *seaweed    E. coli*

| Kingdom | Example organisms |
|---------|-------------------|
| Animal  |                   |
| Plant   |                   |
| Microbe |                   |

    **c** Describe **one** way in which organisms within the animal kingdom can be further classified.

    **d** Describe **two** benefits of scientists classifying organisms.

**2** List **three** factors that are needed for the survival of:

    **a** plants

    **b** animals.

**3** Animals and plants are adapted to live successfully in their environments.

    **a** What is meant by an adaptation?

    **b** Name a plant that is adapted to live in the desert.

    **c** List **four** adaptations for this plant.

    **d** Explain how each adaptation allows the plant to survive in the desert.

**4** Sort the statements below into the correct order, to describe the process of natural selection:

Genes from successful organisms are passed on to their offspring

This process is repeated many times

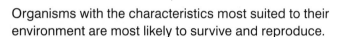

Organisms in a species show variation, caused by genetic differences

Organisms with the characteristics most suited to their environment are most likely to survive and reproduce.

**5** Give **three** examples of how plant hormones can be used to help crops grow effectively.

**6** Plant growth is affected by a number of different factors.

    **a** Name **two** factors that affect plant growth.

    **b** What chemicals do plants use to respond to their environment?

    **c** Where are these chemicals produced?

    **d** What is the difference between phototropism and gravitropism?

    **e** Draw a labelled diagram to explain how plants grow towards the light.

# AQA Examination-style questions

**1** Acidophiles are a type of extremophile that lives best in very acidic conditions. The table shows the percentage of species that survive in different pH levels.

| pH | Amount of species that survive (%) |
|---|---|
| 1 | 87 |
| 2 | 75 |
| 3 | 70 |
| 4 |  |
| 5 | 49 |

**a** Predict the percentage survival at pH 4. *(1)*

**b** In which pH level do the acidophiles have the best chance of survival? *(1)*

**c** Calculate the percentage increase in the amount of species that survive between pH 2 and pH 1. *(1)*

**2** Choose words from the box to complete the sentences.

Words can be used more than once, once or not at all.

> *compete   food   short   survive   water*

Living things ............ for resources that are in ............ supply, such as ............ and ............ Those plants and animals that ............ successfully will ............ to breed. *(6)*

**3** Auxin is a hormone that controls plant growth.

Use the diagram of a root to help you answer the questions.

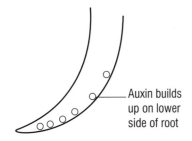

Auxin builds up on lower side of root

**a** How does auxin cause roots to grow downwards? *(3)*

**b** What role do roots play in the plant's survival? *(1)*

**4** Dandelions have many adaptions that allow them to survive.

Match each adaption to how it helps the dandelion to survive.

| Adaption | How it helps |
|---|---|
| Deep roots | Offspring spread far away so not in competition with parent |
| Grows quickly on bare soil | Chemical methods will not kill it |
| Leaves spread out over ground | No low lying plants can get sunlight |
| Resistant to many weedkillers | Collect more water |
| Seeds are spread by the wind | Can grow anywhere |

*(4)*

**5** A penguin and an osprey look very different.

**a** *In this question you will be assessed on using good English, organising information clearly and using specialist terms where appropriate.*

Give reasons for and against scientists classifying them in the same group. *(6)*

**b** Suggest **one** thing that penguins might compete with other animals for. *(1)*

**6** Some seed-eating finches from the mainland colonised the Galapagos Islands thousands of years ago. These had thick beaks to crush seeds. There are many types of finches on the islands now, some with thick beaks, but others have slender beaks to crush insects.

**a** Why would a slender beak be an evolutionary advantage on the islands? *(1)*

**b** Suggest what might have happened to the population of seed-eating finches on the islands since their colonisation. *(2)*

**c** Explain the term '**natural selection**'. *(2)*

## 4.1

# Biomass and food chains

### Learning objectives

- What is biomass?
- What are producers and consumers?
- What is a food chain?

In order to understand the natural world scientists need to describe the flow of energy through the environment. Energy is passed from organism to organism via feeding. This knowledge helps us understand the impact of human activities on the environment and help with conservation efforts.

## Biomass, producers and consumers

**Biomass** is the name given to the mass of all living material found in an **ecosystem**. An ecosystem is all the living matter (such as plants and animals) and non-living matter (such as rocks and water) in an area. For example, the biomass in a pond ecosystem would include animals (such as frogs, fish and insects), plants and algae.

Living organisms can be divided into two groups:

**producers** – those who make their own food

**consumers** – those who cannot.

**a** What is the difference between a producer and a consumer?

## Producers

Producers make their own food through the process of **photosynthesis**. They include all plants and algae.

The **biosphere** is the part of the Earth and the atmosphere where living organisms can survive. Energy enters the biosphere from the Sun as light energy. Producers absorb this light energy and transfer it into chemical energy. They then store it in organic compounds, such as **carbohydrates**, which can then be converted further into sugars, fats and proteins. These are used for growth, repair and as a source of energy. Ultimately they will provide energy for other organisms present in the **ecosystem**.

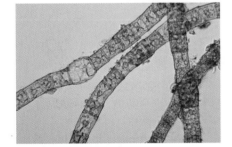

**Figure 1** Algae are simple organisms that produce their own food by photosynthesis

Photosynthesis can be summarised by the following equation:

carbon dioxide + water $\xrightarrow[\text{light energy (from the Sun)}]{}$ glucose (sugar) + oxygen
(contains chemical energy)

$$6CO_2 + 6H_2O \longrightarrow C_6H_{12}O_6 + 6O_2$$

**b** Where do producers get their energy from?

## Consumers

All animals are consumers. They cannot make their own food, and so they have to eat other organisms to gain energy. When an organism is eaten, biomass is transferred to the consumer.

**Decomposers** are also a type of consumer. They gain their energy by feeding on dead or decaying material.

⬭⬭ **links**

*For more information about decomposers and their role in nutrient cycling see 4.4 Recycling of nutrients.*

Consumers gain energy from their food (biomass) through the process of **respiration**. Respiration takes place inside an organism's cells. It can be summarised by the following equation:

$$\text{glucose} \quad + \quad \text{oxygen} \quad \longrightarrow \quad \text{carbon dioxide} \quad + \quad \text{water} \ (+ \text{energy})$$

$$C_6H_{12}O_6 \quad + \quad 6O_2 \quad \longrightarrow \quad 6CO_2 \quad + \quad H_2O \ (+ \text{energy})$$

## Food chains

A food chain displays what organisms eat. The arrows in a food chain show the movement of energy (stored in food) from one organism to the next. Each step in the food chain is known as a **trophic level**. An example of a simple food chain is shown below:

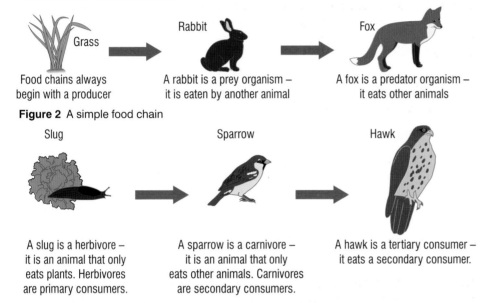

| Grass | Rabbit | Fox |
|---|---|---|
| Food chains always begin with a producer | A rabbit is a prey organism – it is eaten by another animal | A fox is a predator organism – it eats other animals |

**Figure 2** A simple food chain

| Slug | Sparrow | Hawk |
|---|---|---|
| A slug is a herbivore – it is an animal that only eats plants. Herbivores are primary consumers. | A sparrow is a carnivore – it is an animal that only eats other animals. Carnivores are secondary consumers. | A hawk is a tertiary consumer – it eats a secondary consumer. |

**Figure 4** Consumers are further classified to determine their position in a food chain

**c** What is the difference between a herbivore and a carnivore?

In most ecosystems, animals will eat more than one type of organism. For example, a sparrow will also eat seeds and worms. To illustrate this, **ecologists** (scientists who look at how ecosystems work) draw **food webs**. These contain a series of interlinked food chains.

**Figure 3** A lion is an example of a top predator. It is not eaten by other organisms.

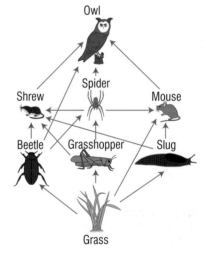

**Figure 5** A food web

## Summary questions

1 Match the following definitions to their meanings:

   Biomass    An organism that makes its own food

   Producer   An organism that eats other organisms to gain energy

   Consumer   All the living matter present in an area

2 Rearrange the following organisms into a food chain:

   grasshopper   frog   grass   snake

3 Using the food chain below answer the following questions:

   cabbage $\longrightarrow$ rabbit $\longrightarrow$ fox

   If all of the rabbits died, what would happen to:
   **a** the number of foxes?       **b** the number of cabbages?

## Key points

- Biomass is the name given to all the living organic matter present in an area.
- Producers make their own food by photosynthesis. Consumers have to eat food to gain energy.
- Food chains and webs show the flow of energy and biomass between organisms in an ecosystem.

## 4.2

# Energy transfer within food chains

### Learning objectives

- How much energy is transferred between organisms?
- How is energy wasted during photosynthesis?
- How is energy lost from a food chain?

### Energy conversion and transfer

The energy contained in food originally comes from the Sun in the form of light energy. This is then transferred to chemical energy by producers, during photosynthesis.

Energy is transferred when one organism eats another. However, not all of the energy is transferred from one organism to the next – energy is wasted in a number of different ways. Eventually, all of the energy will be transferred from the biosphere, heating the surroundings.

### Energy transfer through producers

Not all of the light energy coming from the Sun is transferred into chemical energy by photosynthesis. Only about 1 per cent of the energy a plant receives is transferred. This is because:

- some of the light is reflected from the leaf back into the atmosphere
- some light passes through the leaf
- not all the light is of the correct wavelength for the plant to use
- some energy is used in photosynthesis reactions.

The chemical energy gained is used by the plant for growth. This increases a plant's biomass and provides food for consumers.

**a** What process transfers light energy to chemical energy?

### Energy transfer through consumers

When one organism eats another, only around 10 per cent of the energy is transferred. Therefore, at each trophic level, less and less energy is available. This means that food chains do not consist of many stages. Most have only four levels.

**AQA** *Examiner's tip*

The percentage of energy transfer will never be more than 100 per cent. If your calculation gives an answer over 100 per cent, you have made a mistake, so check your working.

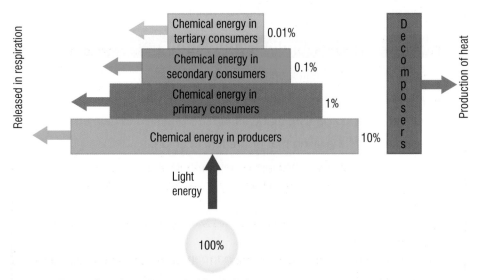

**Figure 1** Energy flow through a typical food chain

Energy might not be transferred between organisms because:

- Some parts of a plant or animal might not be eaten. For example, leaves may be lost from trees. These may then be broken down by decomposers.
- Some parts of the plant or animal cannot be digested and these will be lost from the body in faeces.
- Energy released by respiration is used for movement and other body processes. It is eventually transferred to energy heating the surroundings.
- Energy is transferred from the body in urine – waste products.

**b** Approximately how much energy is transferred from one trophic level to the next?

### Maths skills

#### Energy efficiency in a food chain

Figure 2 shows that the chicken has taken in 110 kJ of energy from its food. Of this, 66 kJ are lost through waste, and 32 kJ are lost through respiration. How much energy would be available to a fox that ate this chicken?

To calculate how much energy is transferred from one organism to the next, you can use the following equation:

$$\begin{array}{ccc}\text{energy} \\ \text{transferred}\end{array} = \begin{array}{c}\text{energy} \\ \text{taken in}\end{array} - \begin{array}{c}\text{energy transferred} \\ \text{in waste}\end{array} - \begin{array}{c}\text{energy transferred} \\ \text{through respiration}\end{array}$$

$$= 110\,\text{kJ} - 66\,\text{kJ} - 32\,\text{kJ}$$

$$= 12\,\text{kJ}$$

Or, expressed as a percentage:

$$\% \text{ energy transferred} = \frac{\text{Energy transferred}}{\text{Total energy taken in}} \times 100 = \frac{12}{110} \times 100$$

$$= 10.9\%$$

**c** How does a chicken transfer energy that is not passed along a food chain?

### Summary questions

**1** Copy and complete using these words:

*respiration   chains   energy   ten   waste*

Food ............ show how energy is transferred between organisms. At each trophic level, less ............ is available – only ............ per cent, approximately, is transferred between organisms. Energy is 'wasted' through ............ products and ............ .

**2** Using the table below, answer the following questions:

| Energy in: | Food (eaten by organism) | Biomass (contained in organism) | Faeces and waste (lost from organism) | Respiration (released as energy) |
|---|---|---|---|---|
| Energy (kJ) | 150 | 20 | 120 | ? |

**a** How much energy is used in respiration and given out as heat to the surroundings?

**b** How efficient is the energy transfer in this organism, expressed as a percentage?

### Did you know … ?

Although less energy is available at each trophic level, energy is never destroyed. Energy may be transferred to a different and possibly less useful form, such as heating up the organism and its surroundings. This energy is no longer available to be passed on to the next organism in the chain. Energy is also transferred to the environment from the food chain through faeces. This energy is not passed on but can be used by other organisms that eat the faeces, such as dung beetles.

Chicken ? kJ biomass

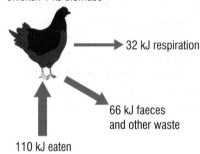

→ 32 kJ respiration

66 kJ faeces and other waste

110 kJ eaten

**Figure 2** Energy transfer in a chicken

### Key points

- Approximately 10 per cent of energy is transferred from one level of the food chain to the next. The remaining energy is transferred, heating organisms and their surroundings eventually.

- Producers do not convert all of the Sun's light energy into chemical energy. Some is reflected and some is not of the correct wavelength.

- Consumers do not pass all of their energy onto the next organism. Some is lost through respiration and waste and some parts of the organism may not be eaten.

# 4.3 Pyramids of numbers and biomass

## Learning objectives

- What is a pyramid of numbers?
- What is a pyramid of biomass?
- How can you draw a pyramid of biomass?

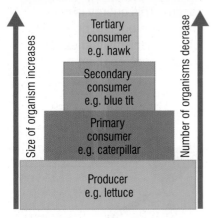

**Figure 1** A typical pyramid of numbers. As you move from one trophic level to the next, the size of organisms generally increases. However, there are fewer and fewer organisms at each level.

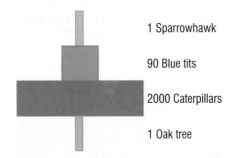

**Figure 2** An inverted pyramid of numbers

## Pyramids of numbers

Food chains show the flow of energy and biomass through a community. However, they do not show how many organisms are involved, or how much biomass is being transferred.

A pyramid of numbers is used to show the population at each level in the food chain. The producer in the food chain is placed at the base of the pyramid. The width of each bar in the pyramid represents the number of organisms present.

The diagram usually has the shape of a pyramid. An organism normally eats more than one organism in the trophic level below. For example, a blue tit needs to eat many caterpillars to survive.

> **a** What generally happens to the number of organisms as you move up a food chain?
>
> **b** What generally happens to the size of the organisms as you move up a food chain?

## Inverted pyramids of numbers

Not all pyramids of numbers are pyramid shaped. This is because these diagrams do not take into account the *size* of the organisms present at each trophic level.

For example, a single tree can support a large number of living organisms. Therefore, when a tree is the producer in a food chain it will result in an inverted pyramid of numbers.

> **c** Why is the bar for an oak tree smaller than the one representing caterpillars?

## Pyramids of biomass

Many scientists choose to represent population data in a pyramid of *biomass*. Here the biomass of the living organisms at each trophic level is calculated. This takes into account both the *number* and *size* of the organisms present at each trophic level. Pyramids of biomass are *never* inverted. Figure 3 shows the oak tree pyramid of numbers, as a pyramid of biomass.

> **d** How is a pyramid of biomass different from a pyramid of numbers?

## Calculating biomass

To collect data to construct a pyramid of biomass, scientists:

- take samples of organisms from each trophic level
- measure the average mass of each of these organisms
- use this data to calculate the total biomass at each trophic level.

Scientists normally calculate the dry mass of an organism, as water content can vary between individuals. This often requires the organisms to be killed and dried in a kiln.

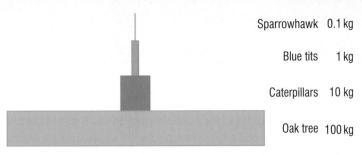

Sparrowhawk 0.1 kg

Blue tits 1 kg

Caterpillars 10 kg

Oak tree 100 kg

**Figure 3** A pyramid of biomass from the pyramid of numbers in Figure 2

| Organism | Number present | Mass of one organism (kg) | Total biomass (kg) |
|---|---|---|---|
| Oak tree | 1 | 100 | 100 |
| Caterpillar | 2500 | 0.004 | 10 |
| Blue tit | 50 | 0.02 | ? |
| Sparrowhawk | 1 | 0.10 | 0.10 |

The total biomass contained in a trophic level can be calculated from:

Total biomass = Mass of one organism × Number of organisms

e.g. Biomass of blue tits = Mass of one blue tit × Number present

= 0.02 × 50 = 1 kg

## Drawing a pyramid of biomass

Pyramids of biomass are scale diagrams. The width of a bar in the diagram represents the biomass of organisms in the trophic level. Using the data from the table above, a sensible scale would be 1 cm = 10 kg. The width of each bar can then be calculated by dividing each biomass by the scale factor (10 kg/cm in this case). For example:

Oak tree: $\dfrac{100\,kg}{10} = 10\,cm$     Blue tit: $\dfrac{1\,kg}{10} = 0.1\,cm = 1\,mm$

**e** Create a scale pyramid of biomass for the food chain:
Oak tree ⟶ Caterpillar ⟶ Blue tit ⟶ Sparrowhawk

### Summary questions

1 Copy and complete using the following words:

*trophic   inverted   biomass   numbers   size*

Pyramids of ............ represent the number of organisms at each ............ level. However, they can sometimes be ............ .

Pyramids of ............ take into account both the number and ............ of organisms present at each level.

2 **a** Complete the table below. Then draw its pyramid of biomass.

| Organism | Number present | Mass of one organism (kg) | Total biomass (kg) |
|---|---|---|---|
| Rose bush | 1 | 4 | |
| Aphid | 2000 | 0.0001 | |
| Ladybird | 5 | 0.002 | |

**b** What would a pyramid of numbers look like for this food chain?

**Figure 4** The sparrowhawk is at the top of a pyramid of numbers or biomass

### Key points

- Pyramids of numbers represent the number of organisms present at each trophic level.

- Pyramids of biomass represent the total amount of biomass present at each trophic level.

- To calculate the size of each bar in a pyramid of biomass, multiply the average mass of the organism by the number of organisms present at that trophic level.

# 4.4 Recycling of nutrients

**Figure 1** A dung beetle is an example of a detritivore. It helps to break down organic waste.

## What are decomposers and detritivores?

Imagine what the world would be like if plants and animals didn't decay when they died. Not only would it soon be swamped with waste, but all the nutrients within that waste would never be available for use.

Thankfully, this doesn't happen. When plants and animals die, their bodies are broken down by **decomposers.** This releases the elements they contain back into the environment. These elements can be used again when they are absorbed by plants in the form of soluble mineral salts. For example, nitrogen is obtained from nitrates.

Decomposers are microorganisms – bacteria and fungi (moulds) – which break down dead organic material. They also break down animal waste – faeces and urine.

**Detritivores** are small animals that help to break down organic material. This speeds up decomposition. They shred the dead material into very small pieces, which makes it easier for decomposers to break down. Examples include:

- earthworms – break down dead leaves
- woodlice – break down wood
- maggots – break down animal material.

   **a** What is the difference between a decomposer and a detritivore?

## How do decomposers release nutrients?

Bacteria and fungi have **enzymes** that break down complex chemicals in organic matter. They produce soluble nutrients, which they can then absorb into their bodies. The nutrients are used for growth and energy. They include amino acids and sugars. If the bacteria and fungi are then eaten by other organisms the nutrients are passed on. In addition, some of the nutrients are released directly into the environment.

## What are the optimum conditions for decomposition?

Microorganisms decompose materials most efficiently in conditions that are:

- **Warm** – at high temperatures, the enzymes used by microorganisms are destroyed. At low temperatures, the process of decay is slowed down by reducing the enzymes' rates of reaction.
- **Moist** – if not enough water is available, reactions within the microorganisms will slow down or be prevented altogether.
- **Oxygen-rich** – oxygen is needed for the microorganisms to respire. Anaerobic (oxygen-free) conditions will prevent most forms of decay.

   **b** What are the optimum conditions for decay to take place?

## Nutrient cycling within an ecosystem

Plants need nutrients (minerals, elements and organic compounds) for growth. They obtain these from the soil. These are then passed on to animals when the plant is eaten. When plants shed material, such as leaves, and when plants and animals die, decomposers release the nutrients trapped in them. Many of the nutrients find their way back into the soil, where they are absorbed by plants.

This process is known as nutrient cycling. It is summarised in Figure 2.

**c** Where do animals get their nutrients from?

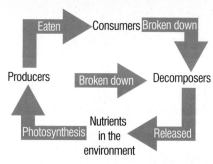

**Figure 2** Nutrients are constantly cycled through an ecosystem

---

### Activity

#### Make your own compost

Compost is nutrient-rich decayed organic material. It provides an excellent growing medium for plants and composting is a very good way of recycling waste. Instead of throwing away kitchen scraps and garden waste, try making your own compost. Follow these four simple steps.

**1** Collect kitchen scraps, such as fruit and vegetable peelings, egg shells, tea bags and toilet roll tubes. Do not use the remains of meat or fish or cooked food.

**2** Add the kitchen scraps to garden waste (grass clippings, etc.), in approximately equal amounts, to your compost bin. (A normal plastic dustbin with air holes drilled into it can be used.)

**3** Wait! It takes between 9 and 12 months to become compost, but during this time you can keep adding to the bin.

**4** Once the compost has turned into a crumbly dark material, which resembles thick moist soil, it is ready to use.

#### Questions

**a** Why is compost good for your garden?

**b** Name at least **two** advantages of making your own compost, compared with buying compost from a garden centre.

**c** Are there any disadvantages of making your own compost?

**d** Why do some people add worms to their compost bins?

**e** Why do compost bins often feel warm?

---

### Summary questions

**1** Copy and complete using the following words:

*small   bacteria   detritivores   decomposers   dead*

Microorganisms called ............ return nutrients to the environment by breaking down ............organic material. They include ............ and fungi. Small animals called ............ speed up decomposition by shredding the organic material into ............ pieces.

**2** How do decomposers break down organic material?

**3** Why would vegetables keep longer in an airtight container in the fridge, rather than in a fruit bowl on the kitchen table?

---

### Key points

- Decomposers are microorganisms that break down dead organic material.

- Detritivores are small animals that speed up decomposition by shredding organic material into smaller pieces.

- Nutrients are constantly cycled through the ecosystem, so they can be used over and over again.

- We can recycle organic waste from gardens and kitchens.

<table>
<tr><td>

**4.5**

</td><td>

# The carbon cycle

</td></tr>
</table>

## Learning objectives

- What is the carbon cycle?
- How is carbon removed from the atmosphere?
- How is carbon released back into the atmosphere?

## Why is carbon important?

All living organisms need the element carbon to survive. Carbon is used to make carbohydrates, fats and proteins – the building blocks of life. These nutrients are essential for growth and repair. Carbon is constantly recycled through the environment in a number of different forms.

**a** Name **three** molecules containing carbon that are used by living organisms.

The **carbon cycle** is the series of processes by which carbon circulates through the environment. It passes through the atmosphere, and into and out of the Earth, and living organisms. It can be summarised in the diagram shown below.

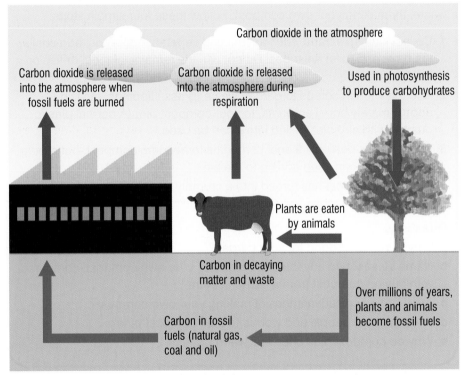

**Figure 1** The carbon cycle. The red arrows show where carbon is absorbed by living organisms. The blue arrows show where carbon is released from living organisms.

## How carbon is removed from the atmosphere

Carbon dioxide ($CO_2$) is removed from the environment by green plants and algae. During photosynthesis they use light energy from the Sun to convert carbon dioxide and water into glucose and oxygen.

Glucose is a simple sugar. It can be used to make complex carbohydrates, such as starch, fats and proteins, which are all needed by plants to grow and develop. This turns the carbon into extra biomass.

When plants are eaten by animals, carbon in the plants is transferred to the animals. Some of this carbon is used to produce fats and proteins in the animals' bodies.

## links

*For more information about photosynthesis and respiration look back at 4.1 Biomass and food chains.*

## Practical

### The test for carbon dioxide

In the laboratory, try breathing out gently through a straw into a solution of limewater. If it turns cloudy, it shows that you are breathing out carbon dioxide. This is produced when your body respires to release energy from your food.

● How would you show that the air you breathe out contains more carbon dioxide than the air you breathe in?

**Safety:** Wear eye protection.

**b** How do animals obtain carbon?

## How is carbon released back into the atmosphere?

There are three main ways carbon is released back into the atmosphere.

**1 Respiration** – plants, algae and animals respire to release energy from their food. Carbon dioxide is produced as a result of the chemical reactions that take place during respiration. It is released back into the atmosphere.

**2 Decomposition** – when plants, algae and animals die, decomposers break down their remains. As the decomposers respire they release carbon dioxide back into the atmosphere.

**3 Burning fuels** – fossil fuels, in particular, are a store of carbon. When they are burned, this trapped carbon is released back into the atmosphere as carbon dioxide. Fossil fuels include coal, oil and natural gas.

**c** Name three examples of fossil fuels.

**Figure 2** In the right conditions, remains of dead plants and animals can, over millions of years, be converted into fossil fuels. This forms a store of carbon until the fossil fuels are burned.

## Summary questions

**1** Copy and complete using the following words:

*photosynthesise dioxide cycle fossil carbon respire animals decompose*

The carbon ............ shows the movement of ............ throughout the biosphere. Carbon ............ is removed from the atmosphere when plants ............ It is then transferred to ............ when they eat these plants. Carbon dioxide is returned to the atmosphere when organisms ............ , when organisms die and ............ , or when ............ fuels are burned.

**2** Name two ways in which carbon can be released back into the atmosphere.

### AQA Examiner's tip

It is useful to construct a flowchart of the carbon cycling between organisms and the atmosphere. Make sure you can name each of the processes involved.

### Key points

● The carbon cycle shows how carbon is constantly cycled throughout the biosphere.

● Carbon is removed from the atmosphere when plants and algae convert carbon dioxide into sugars during photosynthesis.

● Carbon is released back into the atmosphere, as carbon dioxide, when living organisms respire and when fossil fuels are burned.

# 4.6 Human influence on the carbon cycle

## Learning objectives

- How is carbon stored?
- How is limestone formed?
- How do humans influence the carbon cycle?

## Carbon in the environment

Natural processes that maintain the levels of carbon dioxide in the atmosphere include photosynthesis, respiration and combustion.

## Stores of carbon

There are four main ways in which carbon is stored away from the atmosphere:

- **In the oceans** – The oceans hold 50 times as much carbon dioxide as the atmosphere. They also absorb a lot of carbon dioxide created by human activities, such as burning fossil fuels.
- **In the bodies of plants and animals** – This carbon is only stored temporarily and is released by decomposers when the organism dies and its body is broken down. Rain forests are particularly vital carbon stores.
- **In fossil fuels** – Oil, coal and natural gas deposits lock up carbon for millions of years. The carbon is released back into the atmosphere in the form of carbon dioxide when the fuel is burned.
- **In rocks** – Carbon can be stored in rocks, such as limestone and chalk, which are composed mainly of calcium carbonate.

   **a**   Name **three** stores of carbon.

## Limestone

Limestone is formed mainly from the remains of marine animals, such as molluscs, coral and plankton. They convert the carbon dissolved in sea water into calcium carbonate to make shells or other hard parts. When these organisms die, their shells are deposited on the sea bed. Over long periods of time this sediment builds up and forms limestone.

**Figure 1** Limestone (calcium carbonate) acts as a store of carbon. It can take millions of years before this carbon is returned to the atmosphere.

**Figure 2** Coral is an example of a limestone-building organism

Carbon may be stored in limestone for millions of years. This is an example of a 'carbon sink'. Eventually, through movements of the Earth, the limestone may become exposed to the air. As a result of chemical weathering or thermal decomposition by volcanic activity, carbon may then be released back into the atmosphere as carbon dioxide.

> **b** What is the chemical name for limestone?
>
> **c** Which parts of marine animals might form limestone?

Many of the marine organisms that are capable of producing limestone live in warm, shallow, tropical waters. However, it is these areas of the oceans that humans have exploited widely. This has endangered the habitats of these creatures, potentially reducing this carbon store.

## Human influence

Human activity has increased the amount of carbon dioxide being released back into the atmosphere. This has mostly occurred in two ways:

- **Burning fossil fuels** – Carbon dioxide is produced when we burn fossil fuels. Due to increases in the human population and a desire for a higher standard of living, more energy is being used. We use lots of electrical appliances. More fossil fuels have been burned to generate electricity. This has increased the amount of carbon dioxide released into the atmosphere.
- **Deforestation** – Large areas of forest have been cleared for farming, to make space for roads and buildings, agriculture and animal grazing, as well as for the timber itself. This means that there are fewer trees to remove carbon dioxide from the atmosphere by photosynthesis. Therefore, more carbon dioxide remains in the atmosphere.

  In many cases the felled trees are burned or left to decompose. This further increases the levels of carbon dioxide in the atmosphere.

> **d** Why does an increasing use of energy lead to an increase in the amount of carbon dioxide in the atmosphere?

**Figure 3** Large-scale deforestation is still taking place in many areas to create space for farming and supply timber for industry/furniture

### Summary questions

1 Copy and complete using the following words:

   *fossil dioxide limestone deforestation bodies carbon millions*

   ............. is stored temporarily in the ............. of plants and animals. However, it can be stored for ............. of years in ............. rocks.

   ............. and the burning of ............. fuels is increasing the level of carbon ............. in the atmosphere.

2 Explain two different ways that burning forests to clear land for farming affects the atmospheric level of carbon dioxide.

3 Explain why an increased standard of living has led to an increase in the amount of carbon dioxide being released.

### Key points

- Carbon may be stored in the bodies of plants and animals, fossil fuels and limestone rock.
- Limestone is formed under the sea when marine animals, such as coral and plankton, die and their shells and skeletons build up on the sea floor. Over millions of years, these deposits turn to rock.
- Human activity can increase the levels of carbon dioxide in the atmosphere through deforestation and the burning of fossil fuels.

# Summary questions

**1** These organisms form a food chain.

  **a** Rearrange the organisms above into a food chain.

  **b** Which organism is a producer?

  **c** What is the difference between a herbivore and a carnivore?

  **d** What could happen to the number of robins if all the owls died?

**2** A rabbit has consumed 140 kJ of energy from its food. Of this, 35 kJ are lost through waste and 85 kJ are lost through respiration.

  **a** How much energy would be available to a fox that ate this rabbit?

  **b** What percentage of the energy that a rabbit consumes is transferred to the fox?

  **c** Name three ways energy is 'lost' from the rabbit, and not transferred to the fox.

**3** This is a typical pyramid of numbers.

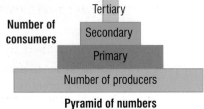

**Pyramid of numbers**

  **a** In general, what happens to the size of the organisms as you move up a food chain?

  **b** In the pyramid shown above, what happens to the number of organisms at each level as you move up the food chain?

  **c** What is the difference between a pyramid of numbers and a pyramid of biomass?

**4** When plants and animals die, their bodies are broken down by **decomposers**. This releases the nutrients they contain back into the environment, where they can be used again.

  **a** Name a type of organism that decomposes materials.

  **b** How do detritivores help decomposition occur more quickly?

  **c** What are the optimum conditions for decomposition?

  **d** Why does food keep fresh for longer in:

    **i** an airtight container?

    **ii** a fridge?

**5** This is a diagram of the carbon cycle.

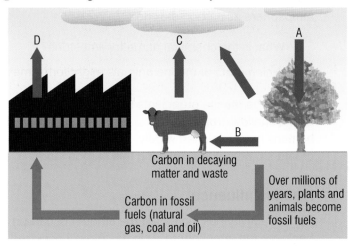

Carbon in decaying matter and waste

Carbon in fossil fuels (natural gas, coal and oil)

Over millions of years, plants and animals become fossil fuels

  **a** Which processes do arrows A to D represent on the carbon cycle?

  **b** How can you test for the presence of carbon dioxide?

  **c** What steps can be taken to limit the amount of carbon dioxide that is being added to the atmosphere through human activity?

**6** Carbon is temporarily stored in parts of the carbon cycle – sometimes for millions of years.

  **a** Calcium carbonate is the chemical name of which rock?

  **b** Describe how calcium carbonate is formed.

  **c** Explain how carbon is released from calcium carbonate, back into the atmosphere.

**7** The pyramid of numbers for a farmland food chain is given below.

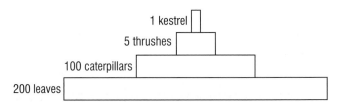

1 kestrel

5 thrushes

100 caterpillars

200 leaves

  **a** Draw and label a pyramid of biomass for the food chain using information from the table.

| Organism | Mass in grams |
| --- | --- |
| Leaf | 5 |
| Caterpillar | 4 |
| Thrush | 70 |
| Kestrel | 250 |

  **b** Calculate the percentage of biomass that is passed from the thrushes to the kestrel.

  **c** Suggest why the biomass of a tree is greater in summer than it is in winter.

# AQA Examination-style questions

**1** Choose the correct words from the box to complete the sentence.

| food fox grass heat movement |

In a food chain of: grass ⟶ rabbit ⟶ fox, all the energy that the rabbit gets from the ............ is not passed on to the ............ as the rabbit transfers some of its energy into ............ and ............ .

*(4)*

**2** Match the correct pyramid of numbers to its food chain.

| Food chain | |
|---|---|
| **a** 1 beech tree ⟶ 50 bark beetles ⟶ 5 blue tits ⟶ 1 sparrowhawk | *(1)* |
| **b** 1 oak tree ⟶ 100 caterpillars ⟶ 10 blue tits | *(1)* |
| **c** 10 lettuce plants ⟶ 2 rabbits ⟶ 100 fleas | *(1)* |
| **d** 300 oak leaves ⟶ 100 caterpillars ⟶ 10 shrews ⟶ 1 owl | *(1)* |

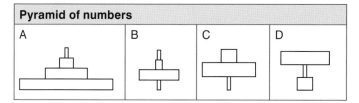

| Pyramid of numbers |
|---|
| A     B     C     D |

**3** The element carbon can move through plants, animals, the air and the ground. This movement is called the carbon cycle.
  **a** Describe the role that plants play in absorbing and storing carbon dioxide. *(3)*
  **b** Name and describe a process by which carbon dioxide is released into the atmosphere. *(2)*

**4** Energy flows through living things in a food chain.

The organisms below are part of a woodland food chain.

| fox     hedgehog     leaves     snail |

  **a** Explain which organism is the producer. *(2)*
  **b** Explain why the fox is called a tertiary consumer. *(2)*
  **c** Suggest how the biomass of a leaf could be found. *(2)*

**5** Energy enters the food chain as sunlight.
  **a** The Sun provides 100 000 kJ/m² per year to producers.
    Of this, plants waste 90 000 kJ/m² per year as heat and 500 kJ/m² per year is inedible.
    Calculate the percentage of energy transferred to the next stage of the food chain. *(4)*
  **b** Explain why not all of the energy in a prey transfers to its predator. *(2)*

**6** The diagram shows the energy transfers of a cow.

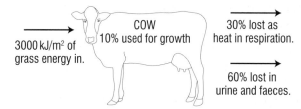

  **a** Calculate how much of the energy that the cow gets per m² of grass is lost. *(2)*
  **b** Use the equation to calculate how efficient the cow is at making biomass.

$$\text{efficiency} = \frac{\text{useful energy}}{\text{total energy}} \times 100\%$$ *(2)*

  **c** Suggest **one** reason why cows spend most of the time grazing. *(1)*

**7** Part of the carbon cycle is shown in the diagram.
  **a** Name the processes labelled A, B and C.

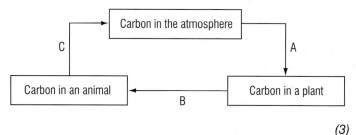

*(3)*

  **b** *In this question you will be assessed on using good English, organising information clearly and using specialist terms where appropriate.*

  Explain how carbon is locked away in the formation of limestone. *(6)*

1  Animals and plants become adapted to their environment.

   **a**  Suggest and explain **two** adaptions of animals to help them live in the desert. *(4)*

   **b**  Suggest and explain an environment where webbed feet would be an advantage. *(2)*

2  Rock salt has to be separated before it is used.

   **a**  Explain why it might be an advantage to crush the rock salt first. *(2)*

   **b**  Explain the process that could be carried out in the lab to separate salt from rock salt. *(6)*

3  Scientists need to know the structure of the atom so they can understand how reactions happen.

   **a**  Nitrogen has a mass number of 14 and an atomic number of 7.
Draw a labelled diagram of a nitrogen atom. *(3)*

   **b**  Explain how an atom can become a positive ion. *(1)*

4  Iron can be separated from its ore using a reduction reaction.

   **a**  Complete the symbol equation for the reduction of iron oxide.

$$Fe_2O_3 + 3CO \longrightarrow \ ......\ Fe + ......\ CO_2$$  **[H]** *(2)*

   **b**  Use the word equation to explain how iron is separated from its ore. *(2)*

5  Materials can be classified as element, compound or mixture.

   **a**  Explain how air could be classified. *(3)*

   **b**  Explain how oxygen molecules could be classified. *(2)*

6  Earth scientists have mapped the structure of the Earth.

   **a**  *In this question you will be assessed on using good English, organising information clearly and using specialist terms where appropriate.*
Describe the structure of the Earth. *(6)*

   **b**  Explain what happens inside the Earth to make a volcano *(4)*

7  Choose the correct word from each box to complete the sentences.

   **a**  Astronomers believe that the universe is ............ .

> **expanding   shrinking   warming**

   **b**  The evidence for this is that light coming from distant stars is ............ .

> **absorbed   squashed   stretched**

   **c**  This change in the light is called ............ shift.

> **blue   green   red**

*(3)*

**8** Magnesium burned in air produces magnesium oxide.

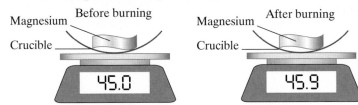

**a** Explain what has happened to the mass of the magnesium. *(2)*

**b** How could the accuracy of the experiment be improved? *(2)*

**9** The level of carbon dioxide has changed a lot over the history of this planet.

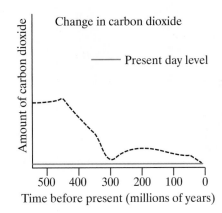

**a** Using the graph suggest the year that animals first started to appear and explain your choice. *(3)*

**b i** Carbon dioxide is a greenhouse gas. Describe the need for greenhouse gases on a planet. *(2)*

**ii** Explain what is happening to the carbon dioxide levels now. *(2)*

# My family

In Unit 2, Theme 1 you will work in the following contexts, covered in Chapters 5 and 6:

## Control of body systems

### How do we react to our environment?

Your nerves help to protect you from danger. Pricking your finger triggers a reflex reaction, causing your hand to be pulled away quickly. A reflex is a very fast response. It prevents further damage to your hand.

Your nerves are also part of the systems the body uses to control your internal environment. For example, they keep your body temperature constant. Nerve receptors in the skin provide information about the external environment. The information is passed to the brain, and changes are made to raise or lower your body temperature as required.

### Can loud sounds damage your hearing?

Sound waves travel outwards from vibrating objects. These are picked up and processed in your ear, so you can hear sounds. Playing music too loudly can permanently damage your hearing. This is a particular problem when using earphones, as sound waves are channelled into the ear.

## Chemistry in action in the body

### How can you stop indigestion?

Hydrochloric acid in your stomach is essential to kill germs and help you digest your food. However, too much acid causes indigestion. You can take 'antacid' tablets to help get rid of the excess acid.

## Why are hazard warning labels used on cleaning products?

Household cleaning products have been developed using acids and bases. Many cleaning products have hazard warning labels on the packaging. A hazard symbol clearly represents the potential dangers of using the product. For example, whether a product is an irritant. The label also explains any safety precautions you should take when using the product.

## Human inheritance and genetic disorders

### What are genes?

Most cells in your body contain a nucleus. Stored inside the nucleus are chromosomes, which carry your genes. Genes are sections of DNA, and are inherited from your parents. The combination of genes you inherit determines your characteristics.

### Can you change the way you look?

Many brothers and sisters look alike. They share similar DNA, resulting in lots of characteristics being the same, or similar. Some of these characteristics are controlled by your genes, and cannot be changed. Others can be altered – for example, hair colour and length. These characteristics are said to show environmental variation.

If there is a family history of a genetically inherited disorder, couples often seek advice from a genetic counsellor. Genetic counsellors calculate the risk of a couple having a child with a disorder, such as cystic fibrosis or haemophilia. These are incurable diseases which can have a large effect on the children and their family. Genetic councillors also offer advice on how to manage the conditions. Due to medical advances in the field of genetics, scientists are now able to screen embryos for the presence of genetic disorders.

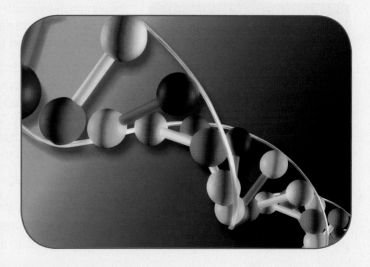

The hormones insulin and glucagon control the amount of sugar that we have in our blood. Some people suffer from diabetes. They are unable to keep their blood sugar at the correct level. Some types of diabetes can be managed through a careful diet and exercise, but some diabetics have to manage their condition through regular injections of insulin.

As you get older, your ears become less sensitive to sound. This applies to both the loudness and the pitch of the sound. The typical human hearing range is between 20 Hz, and 20000 Hz. As your hearing starts to deteriorate, quiet sounds and higher pitched sounds become more difficult to distinguish. With an ageing population, science and engineering have come together to develop a range of hearing aids to help overcome these problems.

## 5.1

# Nerves

## Learning objectives

- What do nerves do?
- What is a controlled nervous reaction?
- What is a reflex action?

Health professionals use their knowledge and understanding of science to treat illness and disease. To do so, they need to understand how our bodies react to internal and external changes.

Your body responds to external changes using your nervous system. The nervous system works by sending electrical impulses around your body. Most information detected by your body is sent to the brain. Your brain is the control centre of your body. The brain processes information received, and decides on an appropriate response. It then sends an impulse to another part of your body telling it what to do.

There are three stages to a nervous response:

1 stimulus (plural, stimuli) – a change in the environment
2 receptors – groups of cells that detect the stimulus
3 effectors – cause a response (they are muscles or glands).

> **a** How are messages sent along nerves?

## Receptor cells

Receptor cells are found in your **sense organs** and are sensitive to a range of stimuli. They change the stimulus into electrical impulses. These travel along nerves (**neurons**) to the **central nervous system (CNS)**. The CNS is made up of the brain and spinal cord.

| Sense organ | Receptor cells | Stimulus |
|---|---|---|
| Eye | Light | Light |
| Ear | Sound | Sound |
| Tongue | Taste | Chemical |
| Skin | Pressure Temperature | Touch Heat |
| Nose | Smell Taste | Chemical |

The table on the left shows where different types of receptor cells are present in your body, and the stimuli they react to.

> **b** Name **four** stimuli that the body responds to.

## Neurons

There are three types of neurons:

- **Sensory neurons** – carry electrical impulses from receptor cells to the CNS.
- **Relay neurons** (found in the CNS) – carry electrical impulses from sensory neurons to motor neurons.
- **Motor neurons** – carry electrical impulses from the CNS to effectors.

### ⬤⬤ links

*For more information on how your body reacts to changes in your internal environment see 5.3 Hormones.*

## Controlled reactions to a stimulus

The flow diagram below shows the steps involved in a controlled nervous reaction:

Stimulus ⟶ Receptor cells ⟶ Sensory neuron ⟶ Spinal cord ⟶ Brain ⟶ Spinal cord ⟶ Motor neuron ⟶ Effector ⟶ Response

This is the way your body normally responds to changes in your environment. The whole process only takes around 0.7 of a second!

### AQA Examiner's tip

It is a common mistake to think that nerve cells are long and thin so that they can travel through the bloodstream more easily. Nerves do not travel through the body but they do stretch over long distances, e.g. from your big toe to your spinal cord.

> **c** What does CNS stand for?
>
> **d** Which type of neuron is only found in the CNS?

# Reflex actions

Reflex actions are automatic reactions that occur without you thinking about them. They do not involve the brain. By missing out the brain, the body can react even more quickly. The body uses these reactions when we are in danger. These may include situations where you could burn or cut yourself.

The flow diagram and Figure 1 below show the steps involved in a reflex action:

Stimulus → Receptor cells → Sensory neuron → Spinal cord → Motor neuron → Effector → Response

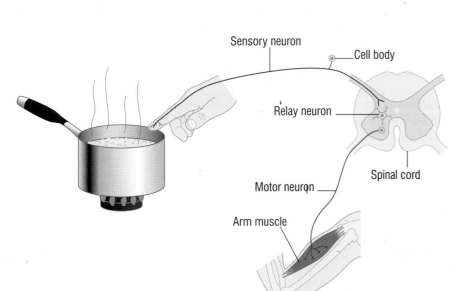

**Figure 1** Reflex actions only take around 0.2 of a second

**e** Which is the fastest type of nervous response?

**f** Describe **three** situations where your body would respond by reflex action.

## Did you know …?

There are approximately 100 billion neurons in the human brain. If we lined up all the neurons in our bodies they would stretch over 1000 km!

## Activity

### Improving reaction times

To win a sprint, athletes need extremely fast reactions to ensure they start running as soon as the gun fires. They improve their reaction rate by practising starting many times. Try to find out if you can improve your own reaction time.

## Summary questions

1 Make a table matching the parts of the nervous system to their function:

| Part of the nervous system | Function |
| --- | --- |
| Receptor cells | Cause a response |
| Effectors | Carry electrical impulses |
| Neurons | Decide on a response |
| Brain | Detects changes in your environment |

2 Are the following reactions controlled or reflex actions?
   **a** Pupils in your eyes shrinking in bright light.
   **b** Tying your shoe lace.
   **c** Signing your name.
   **d** Taking your hand off a hot plate.

3 Draw flow diagrams showing the steps involved in two different situations where a reflex action would be triggered. State the type of receptor cells and the specific effector that would be involved.

## Key points

- Nerves carry electrical impulses around your body, so that your body can react to changes in the environment.

- During a controlled nervous reaction, information about your environment is sent to the brain to be processed. The brain then triggers an appropriate response.

- Your body reacts to danger using a reflex response. These occur without thinking, and occur much quicker than controlled responses.

# 5.2 Hearing ⓚ

## Learning objectives

- How do sound waves travel?
- What is the range of human hearing (in hertz)?
- What is the impact of loud sounds on the environment, society and our health?

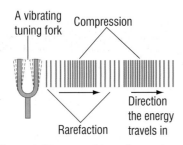

Figure 1 The sound wave is a series of compressions and rarefactions

### AQA Examiner's tip

Make sure that you know the difference between longitudinal waves (sound) and transverse waves (light).

### ?? Did you know ...?

Some shopping centres have installed 'Mosquitoes'. These are speakers that play very high-pitched noises to persuade groups of young people to move on without disturbing older people who cannot hear such high-frequency sound waves. What frequency sound do you think a mosquito device should produce?

## How do you hear a drum playing?

The sound energy travels as a wave from the drum through the air to your ear. When the drum is hit, it vibrates. These vibrations make the air particles (molecules) next to the drum vibrate. The vibrations pass energy on to neighbouring particles, which also vibrate.

- As the vibration passes, some particles are squashed together. This is called a **compression**.
- In other places particles become spread out. This is called a **rarefaction**.

  **a** Sketch a diagram showing how a sound wave travels through air.

Sound waves are one example of **longitudinal waves**. In this type of wave the particles vibrate in the same direction that the energy is travelling in.

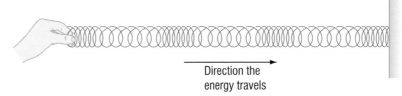

Figure 2 This shows a longitudinal wave travelling along a slinky spring

Vibrating objects cause sound waves. The vibrations can only travel when there are particles that can vibrate, passing the energy on to their neighbours. This means sound travels fast through solids, slowly through gases and not at all in space.

We measure how quickly something vibrates (called its **frequency**) in **hertz**. A vibration of one cycle per second is one hertz. We cannot hear sound waves if objects vibrate at less than 20 hertz. The frequency of the sound is too low for us to hear. The highest pitched sound humans can hear is 20 000 hertz (also called 20 **kilohertz**).

Sounds above 20 000 hertz are called **ultrasound**. As people get older, they cannot hear these very high pitched sounds.

  **b** Write down the frequency of a sound that is too low pitched for us to hear.

Noise levels are measured in decibels. As the decibel reading goes up by 10, the noise level is ten times louder.

The table on the next page shows that a loud rock concert (120 decibels) is 100 times louder than a night club bar (100 decibels). Hearing loss can be caused by any sound above 85 decibels. The damage depends on:

- the loudness of the sounds,
- how long you are exposed to the sounds, and
- how often you hear them.

The table is only a rough guide as sounds are quieter if you stand further away from the source.

| Activity | Sound level in decibels | Maximum exposure time to avoid hearing damage |
|---|---|---|
| Whispering | 20 | No harm |
| Speaking | 50–60 | No harm |
| Loud radio | 65–70 | No harm |
| Busy city traffic | 78–85 | 8 hours |
| Power drill | 95 | 4 hours |
| Night club bar | 100 | 2 hours |
| Road drill | 105 | 1 hour |
| Loud rock concert, MP3 player at maximum volume | 120 | 1 minute |
| Pneumatic drill | 125 | No safe limit |
| Jet engine | 140 | No safe limit |

**c** What affects the amount of hearing damage a person experiences?

The table shows that an MP3 player can be as damaging to your hearing as a loud rock concert. On a bus or train journey, many people listen at high volume to drown out the background noise. The longer you are exposed to the noise, the more damage is likely to occur to your hearing. Specialists are worried that many young people listen to their MP3 players for several hours a day.

The damage to hearing is reversible to begin with. However, after prolonged exposure to loud sounds, people start to suffer from tinnitus (a permanent ringing in the ears). Sometimes they become unable to hear conversations in busy rooms as they cannot hear certain frequencies.

You can prevent this damage by using headphones that fit inside the ear. These can block out the background noise so you can turn your MP3 player down to a safe level.

**d** Why do hearing specialists worry about young people using MP3 players?

By law, employers must not expose their employees to sounds above 87 decibels on a daily basis, or to a peak noise level of 140 decibels. In noisy areas, they must train their staff and provide ear protection to reduce the risk of damage to hearing. They can also sound-proof equipment.

People disturb their neighbours if they cause too much noise. This can include loud music, parties and barking dogs. The noise becomes officially a nuisance at certain decibel levels. Many neighbours manage to sort problems out before calling in the Council. But some cases do reach the courts and people are ordered to stop the noise nuisance.

 **Did you know ... ?**

Astronauts in space communicate by radio as sound waves cannot travel in a vacuum. Victims of earthquakes attract attention more effectively by tapping on pipes rather than shouting.

**Figure 3** Headphones like these can protect your hearing by blocking out background noise so the volume on your MP3 player can be set at a lower level

## Summary questions

1 Complete these sentences by choosing the right word:
   **a** Sound waves travel as **longitudinal/electromagnetic waves**.
   **b** With longitudinal waves, particles vibrate **parallel/at right angles** to the way the energy travels.

2 Explain why sounds cannot travel through space.

3 Research and prepare a poster warning young people of the dangers of loud noise. Your poster can be about one particular aspect, such as concerts, or more general.

## Key points

- Sound travels as longitudinal waves through gases, liquids and solids.

- Humans can hear sounds between 20 hertz and 20 000 hertz.

- Very loud sounds can damage hearing or disturb other people.

# 5.3 Hormones

## Learning objectives

- What are hormones?
- What do hormones do?
- What is homeostasis?
- What is negative feedback? [H]

The body has automatic systems that constantly monitor and control its internal environment. For example, to remain healthy your body must be kept at the correct temperature and have the right amount of sugar in your bloodstream. These processes are controlled by hormones.

Hormones are chemical messengers that travel around your body. They are made in **glands** and secreted into the blood. This carries the hormones all around your body. Hormones cause a response in specific cells found in **target organs**.

Hormones and nerves carry out similar roles but act in very different ways:

|  | Nerves | Hormones |
| --- | --- | --- |
| Speed of response | Very fast | Slower |
| Length of response | Short acting | Longer acting |
| Area targeted | Very precise area | Larger area |
| Time of reaction | Immediate | Longer term |

**a** Where are hormones produced in the body?

**b** What role do hormones play in the body?

The diagram below shows some important glands and the hormones they produce:

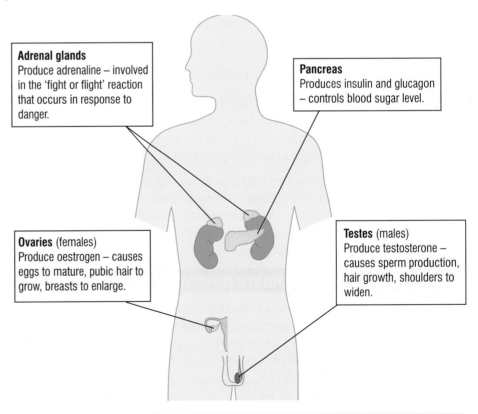

**Adrenal glands**
Produce adrenaline – involved in the 'fight or flight' reaction that occurs in response to danger.

**Pancreas**
Produces insulin and glucagon – controls blood sugar level.

**Ovaries** (females)
Produce oestrogen – causes eggs to mature, pubic hair to grow, breasts to enlarge.

**Testes** (males)
Produce testosterone – causes sperm production, hair growth, shoulders to widen.

**c** Name the **two** hormones that are responsible for changes that take place during puberty.

## What is homeostasis?

Keeping the conditions in your body (your internal environment) constant is called **homeostasis**. This is essential for the body to function normally. It is maintained by the nervous system and hormones.

For example, enzymes require very specific conditions to work effectively. Some enzymes are used to build molecules, which are needed by the body for growth. Others are used to break down molecules – to digest food, for example. Even slight changes in the body's internal environment can slow down or prevent enzymes from working. The process of homeostasis ensures that the optimum conditions are maintained.

> **d** What does homeostasis mean?

## Which body systems are controlled by homeostasis?

Many internal systems have to be controlled. These include:

- **body temperature** – controlled by changes in the skin
- **blood glucose** (sugar) concentrations – controlled by hormones in the liver and pancreas
- **blood carbon dioxide** concentrations – controlled by lungs
- **body water** concentrations – controlled by the kidneys.

*Higher*

## Negative feedback

The body maintains a steady state through the process of **negative feedback**. This means that any changes which affect the body are reversed and returned to normal.

For example, negative feedback is involved in maintaining body temperature:

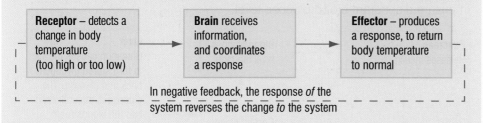

In negative feedback, the response *of* the system reverses the change *to* the system

**AQA** *Examiner's tip*

Remember that your internal conditions are never in a perfect state. They fluctuate about that state using negative feedback to adjust backwards and forwards.

⊂⊃ **links**

*For more information on how the body uses hormones to maintain a constant blood sugar level see 5.4 Diabetes and controlling blood sugar levels.*

*For more information on how body temperature remains constant see 5.5 Controlling body temperature.*

### Key points

- Hormones are secreted by glands. They are chemical messengers that travel around your body in the blood.
- Hormones control body processes that need constant adjustment to maintain a constant internal environment.
- Homeostasis is the maintenance of a constant internal environment.
- The body uses negative feedback to control its internal environment. [H]

## Summary questions

1 Copy and complete using these words:

*blood   glands   homeostasis   target*

Hormones are produced in ............. They travel in the ............., and cause a response in ............. organs. ............. means the maintenance of a constant internal environment.

2 Describe some differences between the way nerves and hormones work.

3 Why is negative feedback important? Illustrate your answer with an example. [H]

# Diabetes and controlling  blood sugar levels

5.4

**Learning objectives**

- What does insulin do?
- What does glucagon do? [H]
- What is diabetes?

Your body needs glucose (sugar) for energy. However, the level of glucose in your blood must be kept constant. Too much glucose in the blood is dangerous, and can cause serious health problems.

## Controlling blood glucose levels

After eating, your blood sugar level rises. Carbohydrates are broken down into glucose (sugar). This causes your blood sugar level to rise. Some of this sugar is used by cells to release energy. Excess glucose is stored in the liver until it is needed.

> **a** What is glucose used for in the body?

The hormones insulin and glucagon are responsible for maintaining a constant blood sugar level. These hormones are both produced by the pancreas.

Insulin is released by the pancreas if blood glucose levels are too high. Insulin makes the liver remove glucose from the blood and store it as insoluble glycogen. This reduces blood glucose to normal levels.

Glucagon is released by the pancreas if blood glucose levels are too low. Glucagon makes the liver convert glycogen back into glucose and release it into the blood. This increases blood glucose back to normal levels.

**Higher**

> **b** Which hormones are responsible for controlling blood glucose levels?

The diagram below shows how insulin and glucagon work together to ensure a constant blood glucose level.

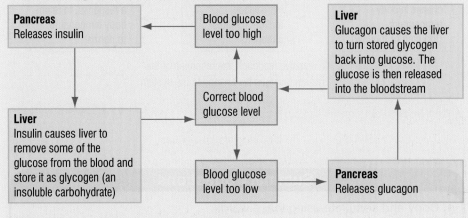

**Figure 1** How blood glucose level is controlled

**AQA** *Examiner's tip*

There are three substances here with quite similar names – glucose, glycogen and glucagon. Make sure that you know the difference between them and practise selecting the correct term.

## What happens if you cannot control your blood glucose level?

If your blood sugar level stays high it can lead to a diabetic coma. If this is not treated you can die. People who cannot control their blood glucose levels suffer from **diabetes**.

> **c** Which disease may people suffer from if they cannot control their blood glucose level?

There are two main types of diabetes:

- **Type 1** – sufferers do not produce insulin. To control their blood glucose level, they have to inject themselves with insulin several times a day. They also have to eat a healthy diet and ensure regular physical activity. This is also known as insulin-dependent diabetes.
- **Type 2** – sufferers do not produce enough insulin, or poor quality insulin is produced. Dieticians advise these people to avoid eating large quantities of carbohydrate-rich foods, and to exercise after they have eaten. This helps use up excess glucose. In severe cases, when controlling diet is not enough, insulin injections are also prescribed.

**d** Why do people with diabetes have to test their blood sugar level before injecting themselves with insulin?

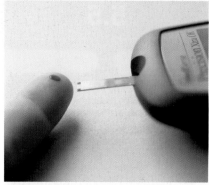

**Figure 2** A blood testing kit, which is used to monitor blood glucose levels

## Activity

### Implications of obesity
### Discussion

Britain has the fastest growing rate of obesity in the developed world. What are the social, economic and health implications for the country if this rate of increase continues?

**Figure 3** Being obese increases your risk of developing Type 2 diabetes by up to ten times

## Summary questions

1  Copy and complete a) and b) using these words:

*glucagon   hormone   increases   decreases   pancreas   glucose*

**a)** Insulin is a ............ produced by the ............ It ............ blood ............ levels.

**b)** ............ is another hormone released by the pancreas. It ............ blood glucose levels. [H]

2  Name some foods that diabetics must avoid eating in large quantities.

3  Explain how glycogen is involved in controlling the amount of glucose in the blood. [H]

4  Explain why eating a bowl of spaghetti affects blood sugar levels and how the body responds to this change. Mention the liver and glycogen in your answer.

## Key points

- Insulin causes the liver to remove glucose from the blood and store it as glycogen. This reduces blood glucose levels.

- Glucagon causes the liver to convert glycogen back into glucose and release it into the blood. This increases blood glucose levels. [H]

- If people cannot control their blood glucose levels they may suffer from diabetes. Type 1 diabetes is controlled using insulin. Type 2 diabetes can be controlled through diet and exercise.

# 5.5 Controlling body temperature

## Learning objectives

- How does your body control its temperature?
- Why do we sweat?
- Why does the amount of blood near the surface of our skin change?

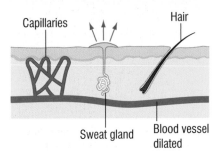

**Figure 1** Skin's appearance when a person is too hot

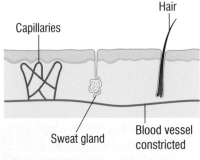

**Figure 2** Skin's appearance when a person is too cold

### AQA Examiner's tip

Just remember what colour your face goes after exercise – bright red. This shows that the capillaries close to the surface of the skin have more blood than normal as the body is trying to cool down. When you are cold your skin looks white as most of your blood is redirected away from the surface of your skin. This is so that you transfer less energy to your surroundings.

Your body works best at 37 °C. Whatever conditions are like outside, your body will try to maintain this temperature.

The thermoregulatory centre in your brain is responsible for controlling body temperature. Your brain monitors the temperature of your blood, and receptors in your skin receive information about the external temperature. The thermoregulatory centre processes this information and sends nerve impulses to the skin and muscles to tell the body how to respond.

**a** What is the normal body temperature?

Sunbathing and physical activity can cause the body to overheat. Exposure to cold weather can cool the body down. Just a couple of degrees difference in your body's temperature can stop the body from working efficiently. This happens especially to the brain. For example, if you are too hot it can cause fits and dehydration.

Your body is designed to protect itself from changes in temperature in several ways, mainly using the skin.

### What happens when you get too hot?

- Hairs on your skin lie flat.
- Sweat glands produce sweat.
- Blood vessels supplying capillaries near the surface of your skin widen (dilate). This increases blood flow through the capillaries, increasing heat loss.

### What happens when you get too cold?

- Hairs on your skin stand on end.
- This traps a layer of air close to the skin, preventing heat loss.
- Sweat glands do not produce sweat.
- Blood vessels supplying capillaries near the surface of your skin narrow (constrict). This reduces blood flow through the capillaries, reducing heat loss.
- Shivering (rapid muscle contractions). This requires extra energy, so your cells respire more. This produces extra heat.

**b** Why do the hairs on your arm stand up when you are cold?

**c** Why does shivering help you warm up?

### Why do you go red when you are hot?

When you are hot the capillaries in your skin widen (vasodilation). This allows more blood to flow close to the surface of the skin and makes you look red. So more energy is transferred from the blood by radiation, cooling you down.

## Why do we sweat?

Sweat is mainly water, but it also contains salt and urea (a waste material). When the water in sweat evaporates it absorbs energy from your body. As energy is lost from your body, your temperature falls so you feel cooler.

The more you sweat, the more you cool down but you also lose more water and salt. These substances must be replaced by drinking and eating, otherwise you will dehydrate.

**d** What is sweat made of?

## What happens if your temperature drops too much?

As your body cools it starts to function at a slower rate. If your temperature drops by 2°C your brain will be affected. Body movements will slow and your speech will begin to slur.

If your body temperature continues to drop, you will go into a coma. Eventually you will die. This condition, when body temperature drops to below 35°C, is called hypothermia. It is a major problem for explorers in extreme weather conditions. It can also affect the elderly if they have poor heating.

**e** How would you know if someone is suffering from hypothermia?

This flow diagram summarises how the thermoregulatory centre in the brain controls body temperature.

**Figure 3** Explorers need to make sure they don't get too cold by wearing protective clothing

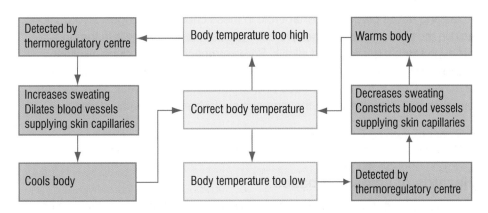

### Summary questions

1 Copy and complete the table with the following words:

*lie flat   make sweat   narrow   stand up   do not make sweat   widen*

| | Body too hot | Body too cold |
|---|---|---|
| **Hairs** | | |
| **Blood vessels** | | |
| **Sweat glands** | | |

2 Explain what the words dilation and constriction of blood vessels mean.

3 If you tasted your skin after you exercised, why would it taste salty?

4 Suggest how a polar bear is adapted so that it doesn't get hypothermia.

### Key points

- The thermoregulatory centre in your brain monitors body temperature. If it is too high or low, it causes changes in the skin that return body temperature to normal.

- The water in sweat evaporates from the surface of the skin. This cools your body.

- When blood flows near the surface of the skin, energy is transferred by radiation to the surroudings. If you are hot, blood flow increases near the surface of your skin increasing the rate of energy transfer. This cools you down.

# 5.6 Body chemistry

## Learning objectives

- How do chemical reactions help our bodies function properly?
- How can acids and bases harm our bodies?
- How can we reduce the risk of harm from acids and bases?

## Useful chemical reactions

Just think about the processes taking place in living organisms:

- **M**ovement
- **R**espiration
- **S**ensing
- **G**rowth
- **R**eproduction
- **E**xcreting waste
- **N**utrition

Each of these processes is controlled by highly specialised chemical reactions. Scientists who study these reactions are called biochemists.

Our bodies are continually breaking chemicals apart and putting them back together in different ways. A good example of this is nutrition. Starches, proteins and fats are all complex molecules we use to feed our bodies every day. They are broken into smaller molecules by enzymes in the body's digestive system. These small molecules are then reassembled to make new materials for the body. For example, proteins are broken down into amino acids. The amino acids are then used to make new, different proteins for the body.

Acids and bases play an important role in these chemical changes. Hydrochloric acid in the stomach helps digestive enzymes to work. Other cells in the body produce soluble bases called alkalis. Some stomach cells produce alkaline mucus (slime) which protects the stomach lining from the hydrochloric acid and enzymes. Cells in the liver produce bile, which contains the weak alkali sodium bicarbonate. This neutralises the stomach acid when it leaves the stomach.

**a** Name **two** organs of the body that produce an alkali.

## Hazards of acids and bases

Acids and alkalis can also have harmful effects on the body. Strong acids are very reactive. They react with other chemicals very easily. If those other chemicals are part of your body, it can be very painful! Acids in the home vary from weak acids, such as vinegar, to stronger acids, such as acids found in toilet cleaners.

As well as burning your skin, acids can damage you on the inside. By cleaning your teeth, you help prevent bacteria growing in your mouth. Bacteria produce acids in their waste. These acids attack your teeth. Stomach ulcers are another example. They are caused by excess stomach acid damaging your stomach lining.

**b** What produces the acid that causes tooth decay?

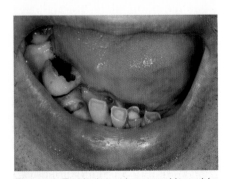

**Figure 1** Tooth decay is caused by acids

Bases can also do a lot of damage to our bodies. If you accidentally get a base on your skin, it will feel slippery. This is because it is turning the oils in your skin into soap-like substances. Stronger bases such as sodium hydroxide (called **caustic soda** in the home) can burn straight through your skin into the flesh below. This is shown in Figure 2.

**c** Why might a base feel slippery if you touch it?

## Staying safe

Any chemical that can harm you should have a label to warn you. These warnings are called hazard symbols. The symbol usually gives an idea of the kind of harm it can cause. The most common hazard symbols you would find on acids and bases are those meaning irritant, harmful or corrosive.

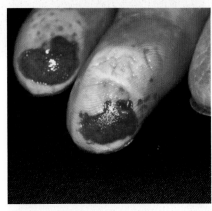

**Figure 2** Bases can be just as dangerous as acids

| <br>IRRITANT | <br>HARMFUL |  |
|---|---|---|
| Irritant – can cause a rash or itching | Harmful – general damage to living organisms | Corrosive – will burn through materials |

As well as reading the label carefully, you should protect yourself with safety equipment. Goggles should always be worn when you work with strong acids or bases. Gloves and masks can be worn when working with harmful chemicals at home. Protective clothing like a lab coat or apron will also give some protection.

**d** What word describes a chemical that causes a rash or itching?

**?? Did you know … ?**

Biologists estimate there are around 75 000 enzymes in the human body. Each of them controls at least one chemical reaction.

---

## Summary questions

1 Copy and complete using these words:

*soap burns ulcers reactions bases gloves protect tooth goggles enzymes*

Body cells carry out many different chemical ............ These are controlled by ............ Acids and ............ can damage your body. Acids can cause skin ............, ............ decay and ............ Bases can turn oils in your skin into ............ Hazard symbols tell us if a chemical is dangerous. We can wear ............, ............ or a mask to ............ ourselves.

2 Name **two** chemicals broken down by your body.

3 Explain why brushing your teeth helps prevent tooth decay.

## Key points

- Chemical reactions control everything that happens in your body.
- Acids and bases are involved in many of your body's chemical reactions.
- Some acids and bases can damage skin, teeth and internal organs.

# 5.7 Acids and bases

## Learning objectives

- Why is the stomach acidic?
- What makes something acidic or alkaline?
- Why do alkalis neutralise acids?

| Name of acid | Formula |
|---|---|
| Hydrochloric acid | HCl |
| Sulfuric acid | $H_2SO_4$ |
| Nitric acid | $HNO_3$ |

Your stomach contains acid, but do you know why, or what type of acid? In fact, your stomach produces between 1 and 3 litres of hydrochloric acid every day. If it didn't, you would find it very difficult to digest food.

The acid in your stomach has two functions:

- It helps digestive enzymes break down protein.
- It kills some bacteria and other microorganisms that might be in your food.

There are lots of ways we can describe acids. We can say they taste sour, or they react with metals, or they turn blue litmus red. The reason acids can do all these things is that they release particles called hydrogen ions ($H^+$).

The table on the left shows the formulae of some common acids. Notice that they all contain at least one hydrogen atom. This is released as a positive ion when the acid is dissolved in water.

We can show hydrochloric acid splitting up to release hydrogen and chloride ions with this equation:

$$HCl \longrightarrow H^+ + Cl^-$$

hydrochloric acid $\longrightarrow$ hydrogen ion + chloride ion

**a** Look at the table on the left. Which type of atom do all these acids contain?

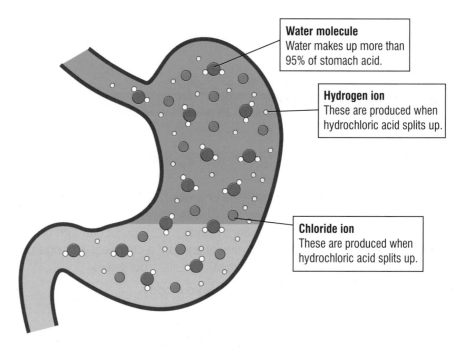

**Water molecule**
Water makes up more than 95% of stomach acid.

**Hydrogen ion**
These are produced when hydrochloric acid splits up.

**Chloride ion**
These are produced when hydrochloric acid splits up.

**Figure 1** Hydrochloric acid in the stomach, containing dissolved $H^+$ ions

The acid in your stomach has to be carefully controlled. If too much is made, it can damage the lining of your stomach. This can cause indigestion and heartburn.

## Bases and alkalis

Luckily, acids can be neutralised by bases. Bases react with the hydrogen ions released by acids. Usually this reaction forms water and a chemical called a salt. We use the word alkali to describe a base that will dissolve in water. Antacids are weak bases that can be used to neutralise stomach acid.

## Measuring acids and bases

We can measure how acidic or basic a chemical is by using the pH scale. The pH scale tells us how much $H^+$ is present. It measures $H^+$ on a scale of 0 to 14. pH 0 means there is a high concentration of $H^+$ ions and pH 14 means there are few $H^+$ ions present. Pure water has a pH of 7: neither acidic nor alkaline. We call this neutral.

Chemicals called indicators can be used to measure pH. They change colour depending on their pH. Different indicators have different colour changes.

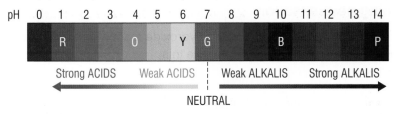

**Figure 2** Universal indicator is a mixture of indicators that helps measure the pH of solutions

**b** What pH would you expect the hydrochloric acid in your stomach to have?

### Practical

#### Determining pH

Your teacher can provide you with a range of chemicals to test in class. You can either use indicators or pH probes to measure how acidic or alkaline each chemical is.

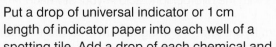

Put a drop of universal indicator or 1 cm length of indicator paper into each well of a spotting tile. Add a drop of each chemical and record the colour change you see. If you are using a pH probe, you can dip the sensor into the chemical you are testing. Make sure you follow any instructions to rinse the sensor after use.

### Summary questions

1 Copy and complete using these words:

*indigestion alkalis neutralise bacteria hydrochloric food*

............ acid is produced in our stomachs. It helps us to digest ............, and kills ............ in our food. Too much acid can cause ............ Antacids are weak ............ that can ............ stomach acid.

2 Name two chemicals that are made when a base reacts with an acid.

3 Which ion causes the acidic solution made when nitric acid dissolves in water?

### links

*For more information on neutralisation see 5.8 Reacting acids with alkalis and bases.*

### AQA Examiner's tip

Make sure you know the difference between an alkali and a base. An alkali is a type of base that can dissolve in water.

### Key points

- The stomach produces hydrochloric acid, which kills bacteria and helps digest proteins.

- Acids all release hydrogen ions ($H^+$) in solution.

## 5.8 Reacting acids with alkalis and bases

### Learning objectives

- Why are carbonates and hydroxides used in antacids?
- What happens in a neutralisation reaction?

**Figure 1** Too much stomach acid can cause indigestion

AQA **Examiner's tip**

Make sure you can write word equations for neutralisation reactions. At Higher Tier you may have to write balanced symbol equations.

| Name of acid | Type of salt made |
|---|---|
| Hydrochloric acid | Chloride |
| Nitric acid | Nitrate |
| Sulfuric acid | Sulfate |

Problems caused by too much acid can often be solved by using alkalis. For example, weak alkalis called antacids can neutralise some of the acid in the stomach. This can help relieve indigestion or heartburn. Alkalis are also useful for neutralising soil or lakes affected by acid rain.

An example of an alkali is sodium hydroxide. Like an acid, it splits up in water to produce two ions:

$$NaOH \longrightarrow Na^+ + OH^-$$

sodium hydroxide $\longrightarrow$ sodium ion + hydroxide ion

The hydroxide ion ($OH^-$) makes the solution alkaline.

### Explaining neutralisation

The hydroxide ions released by sodium hydroxide react with the hydrogen ions from an acid. This produces water like this:

$$H^+(aq) + OH^-(aq) \longrightarrow H_2O(l)$$

The letters in brackets are called state symbols.

(aq) means 'dissolved in water' (or 'aqueous') and the (l) stands for 'liquid'. There are two other state symbols you can use in equations; (s) = solid and (g) = gas.

Any chemical that reacts with the hydrogen ions produced by acids is a base. Other bases include metal oxides and carbonates, as well as metal hydroxides.

**a** Which **two** ions are produced by potassium hydroxide when it splits up in water?

Magnesium hydroxide ($Mg(OH)_2$) is a common antacid that you would use to treat indigestion or heartburn. It reacts with the hydrochloric acid in your stomach.

hydrochloric acid + magnesium hydroxide $\longrightarrow$ magnesium chloride + water

$$2HCl + Mg(OH)_2 \longrightarrow MgCl_2 + 2H_2O$$

As well as producing water, the reaction also makes a salt called magnesium chloride. A salt is always produced when an acid and a base react. In fact, there is a pattern in the types of products made by all neutralisation reactions:

metal hydroxide + acid $\longrightarrow$ metal salt + water

metal oxide + acid $\longrightarrow$ metal salt + water

metal carbonate + acid $\longrightarrow$ metal salt + water + carbon dioxide

The salt that is made depends on which base and which acid are used:

- The first name of the salt comes from the name of the metal in the base.
- The last name of the salt depends on the acid. (See table opposite.)

**b** Which **two** products are always made when an acid reacts with a base?

*Higher*

Here are some examples of antacids neutralising stomach acid. They all follow the rules above:

sodium bicarbonate + hydrochloric acid $\longrightarrow$ sodium chloride + water + carbon dioxide

calcium carbonate + hydrochloric acid $\longrightarrow$ calcium chloride + water + carbon dioxide

aluminium hydroxide + hydrochloric acid $\longrightarrow$ aluminium chloride + water

**c** Which **two** chemicals would you need to make calcium nitrate and water?

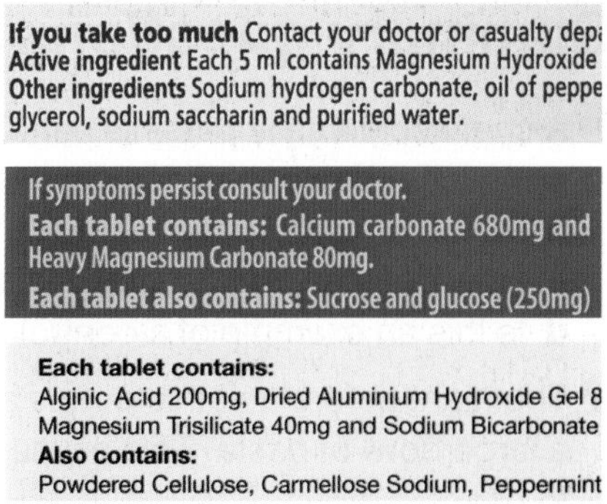

**Figure 2** Antacids are made from carbonates and hydroxides

## Showing neutralisation as balanced symbol equations

Here are the same three reactions shown as balanced symbol equations. For Higher Tier, you must be able to recognise these types of equations and balance them. Remember that in a balanced equation there are the same number and type of reactant atoms as product atoms.

**1** $NaHCO_3 + HCl \longrightarrow NaCl + H_2O + CO_2$
**2** $CaCO_3 + 2HCl \longrightarrow CaCl_2 + H_2O + CO_2$
**3** $Al(OH)_3 + 3HCl \longrightarrow AlCl_3 + 3H_2O$

Balancing these equations also helps to compare the effectiveness of antacids. Notice that aluminium hydroxide can react with three times as much hydrochloric acid compared with sodium bicarbonate.

Different antacids contain different bases and some contain more than one:

| Antacid product | Alkali ingredient | Chemical formula |
|---|---|---|
| Tums | Calcium carbonate | $CaCO_3$ |
| Gaviscon | Aluminium hydroxide | $Al(OH)_3$ |
| Milk of Magnesia | Magnesium hydroxide | $Mg(OH)_2$ |
| Rennie | Calcium carbonate and magnesium carbonate | $CaCO_3$ and $MgCO_3$ |

**d** Which antacids contain carbonates and which contain hydroxides?

### Practical

## Comparing the effectiveness of antacids

Plan and carry out an investigation to see which antacid works best.

● How can you measure the effectivess of each antacid? Let your teacher check your plan before starting the practical work.

## Summary questions

**1** Copy and complete using these words:

*hydrogen   alkaline   salt   carbon   metal*

Hydroxide ions make solutions ............ and neutralise acids. This is because ............ ions react with hydroxide ions. When a ............ hydroxide reacts with an acid, water and a ............ are produced. When a metal carbonate reacts with an acid, water, salt and ............ dioxide are produced.

**2** Copy and complete these word equations:

sodium hydroxide + sulfuric acid $\longrightarrow$

magnesium carbonate + hydrochloric acid $\longrightarrow$

magnesium oxide + hydrochloric acid $\longrightarrow$

**3** Complete and balance the following symbol equation:

$MgCO_3 + HCl \longrightarrow MgCl_2 + ............ + ............$   **[H]**

### Key points

● Hydroxide ions ($OH^-$) make solutions alkaline.

● Bases neutralise acids by reacting with the hydrogen ions that acids produce. They make a salt and water. Carbonates also give off carbon dioxide gas.

**Higher**

# Summary questions

**1** A chef accidentally puts his hand on a hotplate.
  **a** Which receptor cell detects this?
  **b** Which effector is stimulated in this reaction?
  **c** How does an impulse travel along a nerve?
  **d** Is this an example of a controlled or reflex response?

**2** The night before an athlete runs a marathon, she eats a large bowl of pasta. This causes her blood sugar level to rise.
  **a** Why does this occur?
  **b** Which hormone regulates the level of sugar in the blood?
  **c** Where is this hormone made?
  **d** Explain how this hormone returns the blood sugar level to normal?

**3** This picture shows the skin of a builder who has been working hard. What three things in the picture show you that he is hot?

**4** State two features of a sound wave.

**5** Match these symbols to their meanings

  **a** Can cause general damage to people
  **b** Can burn through skin and other materials
  **c** Can cause a rash

**6** Explain why water is one of the products when an acid reacts with an alkali. Use word and balanced symbol equations to help explain your answer. **[H]**

**7** Decide if each of these comments is about hormones, nerves or both. Write your answers in a Venn diagram.

Causes short-term changes in the body

Chemical message

Quick

Responds to stimuli

Travel in the blood

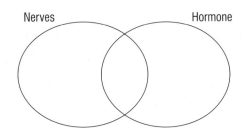

**8** The graph shows the pH of various soaps.

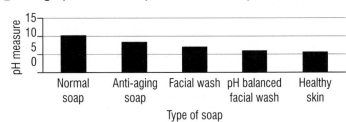

  **a** Which type of soap is nearest to neutral?
  **b** What colour would normal soap turn universal indicator?
  **c** A wasp sting is alkali. Which soap would be best to neutralise the sting?
  **d** Complete the symbol equation by filling in the missing compound.
    $H^+(aq) + OH^-(aq) \longrightarrow$ ............ **[H]**
  **e** Complete this word equation to show a neutralisation reaction.
    hydrochloric acid + sodium hydroxide $\longrightarrow$ ............
    ............ + ............

# AQA Examination-style questions

**1** Label the neurons in this reflex arc. (3)

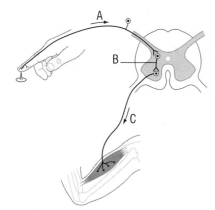

**2** Put these sentences in the correct order to show how your body regulates blood sugar level after eating.
**A** The glucose level in the blood rises.
**B** The glucose level in the blood then falls.
**C** The liver turns glucose into glycogen.
**D** The liver turns glycogen to glucose.
**E** The pancreas releases the hormone insulin.
**F** The pancreas releases the hormone glucagon. **[H]** (5)

**3** A nurse did an experiment to find the best antacid tablet to recommend. The nurse put 10 cm³ of acid in a beaker then added antacid powder one gram at a time. The table shows the pH level after 3 mg of antacid tablet was added.

| Antacid powder | pH after 3 mg of powder |
|---|---|
| Alkalon | 2 |
| Burpeze | 5 |
| Indegon | 4 |
| Tummyset | 4 |
| Windipops | 1 |

**a** Which powder was the most effective? (1)
**b** Predict the pH after 3 mg of Indegon had the nurse used 20 cm³ of acid. (1)
**c** Choose **two** things the nurse should have kept the same to make it a fair test.

| Same beaker | Same amount of acid | Same number of stirs | Same time left in beaker | Same type of acid |
|---|---|---|---|---|

(2)

**4** An example of a neutralisation reaction between an acid and a metal carbonate is given:

HCl + MgCO₃ ⟶

**a** Write a balanced symbol equation for the reaction given. **[H]** (3)
**b** Name the **two** reactants. **[H]** (1)

**5** A recent study has found chocolate to be a benefit to people who are depressed.
**a** Name the organ that releases the hormones that control blood sugar level. (1)
**b** Explain how the body reacts when there is too much sugar in the blood. (3)
**c** *In this question you will be assessed on using good English, organising information clearly and using specialist terms where appropriate.*

Describe the advantages and disadvantages of a person with diabetes eating chocolate. (6)

**6** Body temperature needs to stay constant to within a few degrees otherwise we die.
**a** What is normal body temperature? (1)
**b** Choose words from the box to complete the sentences.

> *air   body   conducting   hair
> insulating   skin*

When we get cold, our ............ stands on end.
This traps a layer of ............ close to the surface
of the ............ Air is ............ so it does not let heat
escape from the ............. (5)
**c** Explain what part the blood vessels close to the surface of the skin have in cooling the body when it is too cold. (3)

**7** Our ears convert sound energy into nerve impulses.

The diagram shows a sound wave travelling from one person to another.

**a** What type of wave is a sound wave? (1)
**b** Which part of the diagram shows a rarefaction? (1)
**c** Explain how the sound wave travels from the girl to the man's ear. (3)
**d** Choose the range of a human's hearing.

| Letter | Range |
|---|---|
| A | Between 0 and 20 Hz |
| B | Between 20 and 200 Hz |
| C | Between 20 and 20 000 Hz |
| D | Between 200 and 20 000 Hz |

(1)

# 6.1 Animal cells ⓚ

## Learning objectives

- What are the main features of an animal cell?
- What are chromosomes and genes?
- How do you inherit features from your parents?

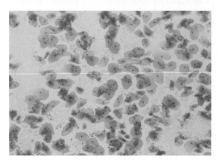

**Figure 1** This is how cheek cells appear when you look down a light microscope. The nucleus is the dark spot in the centre of each cell.

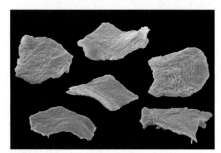

**Figure 2** This is what a nucleus in a cheek cell looks like under an electron microscope. Scientists use electron microscopes to see structures inside cells more clearly.

## AQA Examiner's tip

It may help to remember the parts of a nucleus in decreasing order of size: nucleus, chromosome, DNA, gene.

Your body is made up of millions of **cells**. In fact, cells make up all living organisms. However, they are so small that you can only see them clearly using a microscope. Around 100 animal cells would fit across the width of a full stop! It is hard to work out the exact number of cells in the human body. One estimate is 75 million million!

Scientists use microscopes to look at cells. From this work, we know what happens inside cells. People working in the medical profession can use this information to help fight diseases. For example, scientists view cervical smear test samples under the microscope to look for signs of cancer.

**a** Approximately how many cells are there in the human body?

**b** Look at the two photos on the left. What type of microscope would you use if you wanted to see the smallest features of a cell clearly?

## Animal cells

All animal cells have four main features:

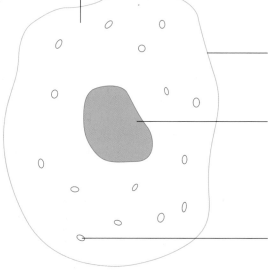

**Cytoplasm** – this is a 'jelly-like' substance. All the chemical reactions in a cell take place here

**Cell membrane** – this is a barrier around the cell. It controls what can come in and out of the cell

**Nucleus** – this contains the information to decide what a cell will look like and what it does. It also contains the information needed to make new cells

**Mitochondrion** (plural **mitochondria**) – this is where respiration happens. Glucose and oxygen react and release energy

**Figure 3** A typical animal cell

**c** Which part of the cell controls the cell's activities?

## What is inside the nucleus?

Inside the nucleus of your cells are chromosomes. These are long strands of the chemical deoxyribonucleic acid or **DNA**, which contain all the information needed to determine the characteristics of a human being.

Each chromosome is divided into sections of DNA. Some of these sections contain the information needed to produce a characteristic, such as eye colour or hair colour. These 'coding' sections are called genes. One chromosome can contain thousands of genes.

**d** What is a chromosome?

**e** What is a gene?

## How do we inherit features from our parents?

We inherit characteristics from our parents through genes. Inside the nucleus of our normal body cells, there are 23 pairs of chromosomes (46 chromosomes altogether). One chromosome of each pair comes from our mother. The other comes from our father.

Egg cells and sperm cells are the only cells in the body to contain just 23 chromosomes. During fertilisation, these cells join together to form an **embryo**. This means that inside each cell of the embryo there are 46 chromosomes (23 pairs).

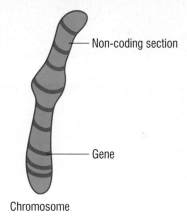

**Figure 4** A chromosome

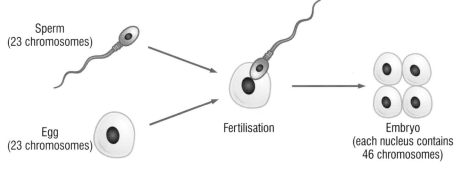

**Figure 5** Fertilisation, followed by cell division

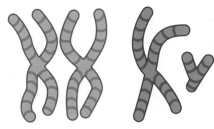

**Figure 6** Out of your 46 chromosomes, 22 pairs are identical, but the 23rd pair (the sex chromosomes) do not always match. They determine whether you are a boy or a girl. Girls have two X chromosomes, whereas boys have one X chromosome and a shorter Y chromosome. Only males carry the Y chromosome, so the chromosomes from the male sperm determine the child's sex.

### Practical

**Making and observing animal-cell slides**

Use a piece of sticky tape to remove some of the dead skin cells from the back of your hand. Add a few drops of methylene blue. Look at the piece of tape through a microscope.

● What cell features can you identify?

### Summary questions

**1** Copy and complete using the words below:

*chromosomes   DNA   genes   nucleus*

Inside each of your normal body cells there is a ............ This contains 46 ............ made up of the chemical ............ Sections of DNA that code for a characteristic are called ............ .

**2** Fruit flies have only four pairs of chromosomes in each nucleus.
  **a** How many chromosomes do they have in a normal body cell?
  **b** How many chromosomes do they have in a sex cell?

**3** Explain why identical twins look the same but brothers and sisters only look similar. (Hint: identical twins occur when a fertilised egg splits.)

### Key points

● Animal cells contain a nucleus, cell membrane, cytoplasm and mitochondria.

● Chromosomes are strands of DNA. On each chromosome the DNA is grouped into sections that code for a specific characteristic. These sections are called genes.

● Parents' characteristics are passed on to their offspring through genes. During fertilisation, genes from the father (contained in the sperm) combine with genes from the mother (found in the egg).

# 6.2

# Variation

**Learning objectives**

● What is variation?

● What can cause variation?

It is easy to tell the difference between a monkey and a fish. This is because they have lots of different **characteristics** (features). However, it is more difficult to tell the difference between two frogs. This is because, within a species, lots of characteristics are shared.

Every person in the world is different – even identical twins are different in some ways. Differences within a species are called **variation**. People vary in many ways including height, build, hair colour and intelligence.

There are two factors that cause variation:

● the characteristics you inherit from your parents – genetic variation

● the environment in which you live – environmental variation.

> **a** What is variation?

## How do people vary?

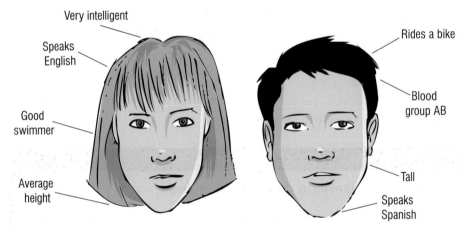

Very intelligent

Speaks English

Good swimmer

Average height

Rides a bike

Blood group AB

Tall

Speaks Spanish

**Figure 1** People can vary in a large number of ways

**Figure 2** Is this person's hair colour inherited?

These two children vary in a number of ways. Some of this variation is due to characteristics they have inherited from their parents. However, most is due to factors in their environment. These include where they live and what they learn from their parents, teachers and friends.

> **b** State **three** characteristics that are caused by genetic variation.
>
> **c** Give **three** characteristics that are influenced by environmental variation.

## Hair colour

This characteristic could be classified as an **inherited** feature. People generally have similar colour hair to one of their parents. However, look at the girl in Figure 2. Do you think her parents also have blue hair? Probably not!

This is an example of environmental variation. The person has chosen to dye and style her hair in this way.

## Height

Height is another characteristic that is mostly determined by your genes. If your parents are tall, you are also likely to be tall. However, if you are very poor with very little to eat, your growth is likely to be stunted as a result of poor diet.

Many characteristics are affected by both environmental and genetic variation, such as height and weight.

> **d** Name **two** other characteristics that are influenced by both environmental and genetic variation.

## Characteristics that are not influenced by the environment

Here are four examples of characteristics that are not influenced by the environment in humans:

- natural eye colour
- natural hair colour
- blood group
- genetic disorders – such as cystic fibrosis and haemophilia.

**Figure 3** These men vary greatly in height

---

### Activity

#### Studying variation

Carry out a survey of some characteristics of the members of your class to see how they vary.

Is the variation in each characteristic a result of environmental factors, genetic factors or a combination of both?

---

### Summary questions

1 Copy and complete the table using the words below:

*weight   intelligence   blood group   skin colour   eye colour hair length   scar*

| Type of variation | Characteristic |
|---|---|
| Genetic variation | |
| Environmental variation | |
| Both types of variation | |

2 **a** Are plants more or less likely than animals to be influenced by environmental variation?
  **b** Name some factors in the environment that could affect plant growth.

3 Why are identical twins the best people to study if you want to find out how the environment influences characteristics?

---

### Key points

- Variation is the name given to the differences that exist between organisms of the same species.

- Variation occurs as a result of genetic factors, environmental factors or a combination of both.

## 6.3 Dominant and recessive alleles

### Learning objectives

- What are alleles?
- What are dominant and recessive alleles?
- How can you work out the chance of a characteristic being inherited?

We have two genes for each characteristic – one from our mother and one from our father. These two genes may be the same. For example, they may both code for blond hair. In this case you will have blond hair.

However, they may be different. Different forms of the same gene are called **alleles**. For example, there is also a gene that codes for brown hair and one that codes for black hair.

**a** What is an allele?

### How do we know which gene will be expressed?

Some genes will always be *expressed* (that is, their characteristic will appear in the individual) – these are called **dominant** alleles. If the gene that codes for black hair is present in the nucleus of your cells, you will have black hair. If the allele for blond hair is also present, it will not be shown. This is because the black hair allele is dominant.

**Recessive** alleles, such as the blond hair allele, will only be expressed if *both* the genes in a pair are recessive. Then you have two of them, and no dominant allele is present .

**b** Which type of allele will always be expressed if it is present?

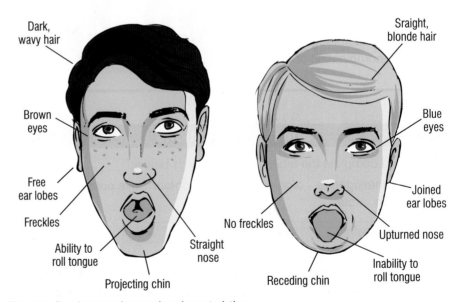

**Dominant characteristics**

Dark, wavy hair
Brown eyes
Free ear lobes
Freckles
Ability to roll tongue
Projecting chin
Straight nose

**Recessive characteristics**

Sraight, blonde hair
Blue eyes
Joined ear lobes
No freckles
Upturned nose
Inability to roll tongue
Receding chin

**Figure 1** Dominant and recessive characteristics

**AQA** *Examiner's tip*

Make sure you understand the terms 'allele', 'dominant' and 'recessive'.

**AQA** *Examiner's tip*

In your exam you may be asked to work out the likelihood of a certain feature being inherited. You should state your answer as a percentage, a fraction or a proportion. For example, on this page the chance of a child having blue eyes is 25 per cent, ¼ or 1 in 4.

**c** Make a list of some recessive characteristics shown in the picture above.

**d** Make a list of some dominant characteristics shown in the picture above.

## How do you inherit eye colour?

To study inheritance, we represent the genes on your chromosomes using letters. The dominant allele is always represented with a capital letter.

To make it simpler to understand, we will only look at the inheritance of one gene. Most characteristics are controlled by several genes. For eye colour, brown eyes are dominant and blue eyes are recessive. In the example below **B** = brown eyes allele, and **b** = blue eyes allele.

During fertilisation, one allele from the mother's egg joins with one allele from the father's sperm. This process can be represented on a Punnett square diagram:

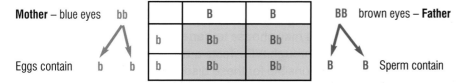

**Figure 2** A Punnett square. These are the possible combinations of alleles from these parents. Studying the inheritance of one gene is called monohybrid inheritance.

All children in this example will have the genes **Bb**. This means that they will all have brown eyes, as **B** is the dominant allele. Remember – you have two genes for each characteristic.

What would happen if both your parents had brown eyes, but they also carried the recessive gene for blue eyes? This is no trickier – just follow the diagram step by step.

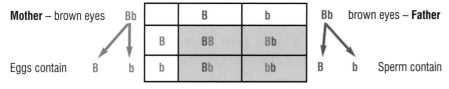

**Figure 3** During fertilisation, any of the combinations shown in the yellow area of the table could be produced

Children would be born in the ratio of 1 **BB** : 2 **Bb** : 1 **bb**. The chances are, for every four children born three will be born with brown eyes and one will be born with blue eyes. Or to put it another way, there is a 25 per cent , or 1 in 4, chance that the child will have blue eyes.

## Summary questions

1 Copy and complete using the words below:

*recessive   alleles   dominant*

Many forms of a gene exist; these are called ............. The ............. alleles will always be shown where present, whereas ............. alleles will only show if a person has two copies of this form of the gene.

2 If you had dominant alleles for hair colour and type – straight or wavy – what would your hair look like?

3 **F** is the gene for freckles and **f** is the gene for no freckles. If a mother has freckles (**Ff**) but the father doesn't (**ff**), what different skin types could their children inherit and in what ratio? Draw a Punnett square diagram like the one above to help explain your answer.

## Key points

● Alleles are different forms of the same gene.

● Dominant alleles are always expressed if they are present in the nucleus. Recessive alleles are only expressed if a person has two copies of this form of the gene.

● Punnett square diagrams are used to work out the chance of a characteristic being inherited.

## 6.4

# Genetically inherited disorders (1)

### Learning objectives

- What are genetically inherited disorders?
- What is cystic fibrosis?
- What is sickle-cell anaemia?

### ?? Did you know ... ?

After fertilisation, a developing baby is referred to as an embryo. After eight weeks of development an unborn baby is then referred to as a fetus.

**Figure 1** Genetic counsellors work out the chance of a couple's child being born with an inherited disorder. They also provide advice and support for people raising children with an inherited disorder.

### Inherited disorders

Many diseases, such as measles and flu occur when particular **harmful** microorganisms enter the body. Other diseases, such as cancer, occur when our bodies go wrong.

Some illnesses, though, are passed down from parents to their children in their genes. These are called **genetically inherited disorders**. Examples include cystic fibrosis and sickle-cell anaemia.

Sometimes, if the risk of having a child with a genetically inherited disorder is high, parents and doctors may choose to **genetically screen** a **fetus** (or embryo, when created through IVF). Tests are performed to detect the presence of the genes responsible for these disorders. If the gene is detected a couple may decide to abort the fetus. Or, in the case of IVF, the embryo will not be implanted.

This is a controversial issue. Some people say this will save the unborn child a life of suffering. However, others believe the unborn child has a 'right to life' regardless.

> **a** What is an inherited disorder?
>
> **b** What is genetic screening?

### Cystic fibrosis

Cystic fibrosis is caused by a 'faulty' gene – a recessive allele (represented here by **c**). A person will only suffer from the disease if both their copies of the gene are 'faulty' (**cc**).

**Symptoms**: Thick sticky mucus is produced. This blocks air passages, making it difficult to breathe. Mucus in the lungs allows germs to grow, causing chest infections. Each infection damages the lungs further, making the sufferer more ill. Excess mucus in the pancreas stops digestive juices being released. This makes it difficult to absorb food.
**Treatment**: At the moment there is no cure. Physiotherapy helps the person cough up the mucus, and antibiotics are used to treat any infections.

People can be **carriers** of cystic fibrosis without knowing it – they have one copy of the faulty gene (**Cc**). They are healthy, but can pass the disorder on to their children if their partner also has a copy of the faulty gene.

| **Mother – Carrier** Cc | | C | c | Cc **Carrier – Father** |

|   | C | c |
|---|---|---|
| C | CC | Cc |
| c | Cc | cc |

Eggs contain C c     C c   Sperm contain

**Figure 2** This diagram shows how two carriers of cystic fibrosis can produce a child who suffers from the disease. There is a 25% (or 1 in 4) chance of the child having cystic fibrosis.

> **c** What are carriers?

## Sickle-cell anaemia

Sickle-cell anaemia is also caused by a recessive allele (represented here by **s**). The allele causes red blood cells to change shape. Instead of being round they become sickle shaped.

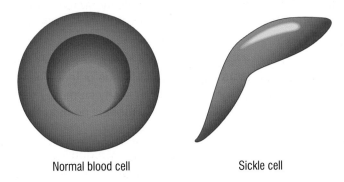

Normal blood cell        Sickle cell

**Figure 3** A normal and a sickle-shaped red blood cell

**Symptoms**: Sickle-shaped red blood cells cannot transport oxygen properly. This results in anaemia. Their shape also means they can become trapped in blood vessels causing pain. People with sickle-cell anaemia are more likely to suffer from infections.

**Treatment**: The condition is managed by pain relief, antibiotics and blood transfusions.

Sufferers used to die very young. Now many survive into their thirties but because most choose not to have children the allele is most commonly passed on through carriers.

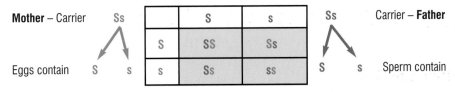

**Mother** – Carrier  Ss

| | S | s |
|---|---|---|
| S | SS | Ss |
| s | Ss | ss |

Ss  Carrier – **Father**

Eggs contain  S  s        S  s  Sperm contain

There is a 25% chance that a child will be born with sickle-cell anaemia.

**Figure 4** This diagram shows how two carriers of sickle-cell anaemia can produce a child who suffers from the disease

### Summary questions

1 Copy and complete using the words below:

*carriers  cystic fibrosis  recessive  sickle-cell anaemia  symptoms*

Two examples of inherited disorders are ............ ............ and ............ ............ . Both are caused by ............ alleles. ............ have one copy of the faulty gene, but have no ............ of the disease.

2 Why would most people be unaware that they are a carrier of a disease?

3 State two arguments for and two against genetic screening. What are your views on the issue of genetic screening?

---

**AQA** *Examiner's tip*

Get plenty of practice drawing Punnett squares so if you are asked for one in an exam you are ready.

---

*Did you know …?*

Being a carrier of the sickle-cell anaemia allele does have one advantage. The allele provides some protection against the deadly disease malaria.

---

### Key points

- Genetically inherited disorders are conditions passed on from parents to their children in their genes.

- Cystic fibrosis is a genetically inherited disorder that is passed on by a recessive allele. Sufferers produce excessive levels of mucus, causing chest infections and difficulty absorbing food.

- Sickle-cell anaemia is a genetically inherited disorder that is also caused by a recessive allele. It makes red blood cells sickle shaped, which stops them carrying oxygen properly.

# 6.5 Genetically inherited disorders (2)

## Learning objectives

- What is polydactyly?
- What is haemophilia?
- How is it hoped that current research will provide cures for genetically inherited disorders?

## Inherited disorders

Genetically inherited disorders are conditions passed on from parents to their children in their genes. They include polydactyly and haemophilia.

## Polydactyly

Some babies are born with extra digits on their hands or feet. These may be fully formed fingers or toes, or no more than fleshy stumps. These extra digits can be removed by surgery but normally there is no medical benefit in removing them.

The most common form of polydactyly is caused by a dominant allele (represented by **P**). A person only needs one copy of the allele to suffer from this disease. People cannot be carriers of this disease – if you have the allele you will have the disease.

> **a** Why is it not possible to be a carrier of polydactyly?

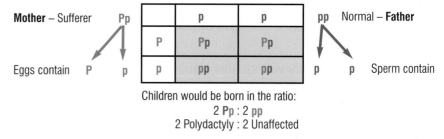

Children would be born in the ratio:
2 **Pp** : 2 **pp**
2 Polydactyly : 2 Unaffected

**Figure 1** A genetic cross between a person suffering from polydactyly and an unaffected healthy individual. There is a 50 per cent chance that their children will suffer from the disorder.

## Haemophilia

Haemophilia is a genetic disorder that only affects males. It is caused by a recessive allele carried on the X chromosome (one of the sex chromosomes).

Females have two X chromosomes. They are carriers of the disease if they have one dominant and one recessive allele. However, they never suffer from the disease, as fertilised eggs containing two recessive alleles will not develop into a baby.

Men have only one X chromosome, so if the 'faulty' allele is present on their single X chromosome, they will suffer from haemophilia.

**Symptoms**: Blood does not clot properly. That is because sufferers cannot produce Factor 8, a chemical that aids clotting, in their blood. So, even small cuts can be dangerous, because they keep on bleeding. Sufferers can lose so much blood that they die. Even small knocks can cause internal bleeding, resulting in very large bruises. Internal bleeding can be severe, causing much pain and damage. If sufferers are not treated they are likely to die before they can have children. Men who have the disorder usually have genetic screening to make sure that the allele is not passed on.

**Treatment**: Regular injections of Factor 8.

## AQA Examiner's tip

When you complete a Punnett square, remember to circle or indicate somehow which is the genotype of the affected offspring.

## ??? Did you know ... ?

The genetic information present in an organism is called its genome. In 2001 the Human Genome Project sequenced the order of the human genome. Scientists are now attempting to identify all the genes in the human genome. It is hoped that the information gained from the human genome project will provide prevention and cures for many diseases.

**b** Which chromosome carries the haemophilia-causing allele?

**c** Why do females not suffer from haemophilia?

## Current research into genetic disorders

Most genetic disorders have no cure. It is hoped that current research may hold the key:

● **Gene therapy**: It may be possible to eventually replace 'faulty' alleles with normal healthy alleles in the tissues where the disease causes damage. For example, to insert healthy alleles into the lung cells of cystic fibrosis sufferers. Alternatively, it may become possible to replace the faulty allele in an embryo.

● **Stem cell** transplants: Embryos contain stem cells. These can grow into any type of cell in the body. Stem cells can be removed from human embryos, or from the umbilical cord after a baby has been born. It may be possible to transplant healthy stem cells into a person who suffers from a genetic disorder.

**Figure 2** Stem cells

---

## Activity

### Stem cell research

Some parents who have a child with an incurable genetic disorder are now choosing to have another child to provide healthy stem cells. The stem cells are taken from the umbilical cord as soon as the sibling is born. These are used to cure their older brother or sister. To ensure the baby does not have the same condition, the embryo would have been genetically screened.

● Research a family who have undergone this treatment. Write a magazine article to explain how and why the family underwent these procedures.

---

## Summary questions

1 Copy and complete using the words below:

*dominant   haemophilia   gene   polydactyly   chromosome*

.......... is a genetically inherited disorder caused by a .......... allele.
.......... only affects males. It is carried on a sex .......... It is hoped that
.......... therapy may be able to cure these diseases in the future.

2 **a** Why is Factor 8 important for the body?
  **b** Can males be carriers of haemophilia? Explain your answer.

3 Draw a Punnett square to determine the likelihood of a child suffering from Huntington's disease, if both parents have the disease – mother (Hh) and father (Hh).

## Key points

● Polydactyly is a genetically inherited disorder, caused by a dominant allele. It results in the production of extra digits.

● Haemophilia is a genetically inherited disorder that prevents blood from clotting properly. It is caused by a recessive allele carried on a sex chromosome.

● Current research into gene therapy and stem cell transplants may be able to cure genetically inherited disorders in the future.

# Summary questions

**1** A student is observing an animal cell through a microscope.

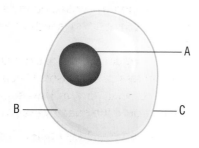

**a** Name parts A, B and C.

**b** Where in the cell do chemical reactions take place?

**c** Which part of the cell contains DNA?

**2** Put these in order of size starting with the smallest.

    **A:** DNA

    **B:** Cell

    **C:** Chromosome

    **D:** Gene

    **E:** Nucleus

**3** Copy and complete the following table. The first one has been completed for you.

| Characteristic | Affected by genes? | Affected by the environment? | Affected by both? |
|---|---|---|---|
| Eye colour | ✓ | | |
| Weight | | | |
| Blood group | | | |
| Skin colour | | | |
| Height | | | |
| Natural hair colour | | | |
| Intelligence | | | |

**4** Many of the characteristics of human beings are affected by their genes.

**a** What is a gene?

**b** What is an allele?

**c** Under which circumstances could a characteristic caused by a recessive allele be expressed?

**5** Cystic fibrosis is a disease caused by a recessive allele. People only suffer from the disorder if they carry both copies of the recessive allele. This allele is given the symbol 'c'. The healthy allele is given the symbol 'C'.

**a** Which alleles would a sufferer have?

**b** Which alleles would a carrier have?

**c** Many carriers do not know that they have the recessive allele. Why is this?

**6** Polydactyly is an example of an inherited disorder. It is a dominant disorder (P).

**a** A man (Pp) and a woman (pp) wish to have a child. Use a Punnett square to show the possible genetic make-up of their offspring.

**b** As a ratio, what is the likelihood of their offspring suffering from polydactyly?

**7** People can change the colour of their hair.

**a** Choose words from the box to complete the sentences about how hair colour is determined.

> *alleles   blond   dominant   genes*
> *hair   inherited   recessive*

We have two ............ for each of our characteristics. Different forms of the same gene are called ............ For example, you might have ............ the gene for black ............ from one parent and the gene for ............ hair from the other. Some genes show up and are called ............ and the gene that does not is called ............ .

# AQA Examination-style questions

**1** The gene for ginger hair is recessive. The diagram shows the inheritance of the ginger gene in a few generations of a family.

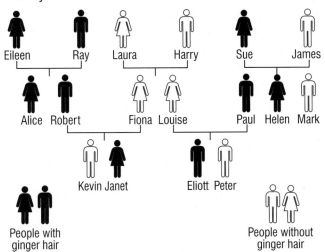

People with ginger hair

People without ginger hair

**a** Which part of the cell carries genetic information? *(1)*

**b** Give the genotype of these people (choose from RR, Rr or rr):
  **i** Sue   **ii** Louise   **iii** Kevin *(3)*

Peter met and had children with someone with ginger hair.

**c i** Fill in a Punnett square to show the possibility of the children inheriting ginger hair. *(2)*

|   | R | r |
|---|---|---|
| r |   |   |
| r |   |   |

  **ii** What is the probability of the children inheriting ginger hair? *(1)*

**2** Choose which of these characteristics are as a result of environmental variation, genetic variation or both.

| Variation | Environmental | Genetic | Both |
|---|---|---|---|
| Foot size |   |   |   |
| Freckles |   |   |   |
| Hand span |   |   |   |
| Scars |   |   |   |

*(4)*

**3** Gardeners win prizes for making new varieties of plant.

A gardener crossed a red flowering sweet pea with a white flowering sweet pea. He collected the seeds and grew them. He noticed that all the seeds grew into red flowering sweet peas.

**a** What colour of sweet pea flower was dominant? *(1)*

**b** Some of these flowers self-fertilised. Calculate the ratio of red and white flowered sweet pea plants he would grow if the seeds were planted. *(3)*

**4** Cystic fibrosis is caused by a faulty gene. The disease affects the lungs by causing an excess of thick, sticky mucus to be made. This mucus traps germs so sufferers often get infections and are constantly on antibiotics.

Scientists have been finding ways of correcting the faulty genes rather than just treating the symptoms. The best way to do it is with gene therapy – replacing faulty genes with normal ones. Trials on mouse and human lung tissue were effective so pilot studies have been started using people who suffer from the disease.

**a i** How would a person first get the disease cystic fibrosis? *(1)*

  **ii** Why would some people be against the gene therapy trials completed so far? *(1)*

**b** Each patient in the pilot study had a small dose of the gene therapy product in the nose every 5 minutes. It took 1.5 hours to give the full dose of 20 ml of the product. Calculate how much gene therapy product was in each dose. *(3)*

**c** *In this question you will be assessed on using good English, organising information clearly and using specialist terms where appropriate.*

Another study was to find the most economical volume of the gene therapy product to use. This would hopefully cure the patient but not cost too much. Plan an investigation that the scientists could carry out to find the smallest volume of therapy product that cures the patients. *(6)*

# My home

In Unit 2, Theme 2 you will work in the following contexts, covered in Chapters 7 and 8:

## Materials used to construct our homes

### Where do materials like concrete come from?

A long time ago, layer upon layer of dead sea creatures built up at the bottom of the sea. Their shells slowly changed into rock. Millions of years later, humans came along and started mining it out of the ground. We call this rock limestone and it is a vital part of the construction industry. Without limestone, we wouldn't have concrete, or mortar to hold bricks together. Limestone is also used in making glass.

### What other materials are used to build our homes?

Metals are also important for making your home work properly. They are used to support concrete, make window frames, wiring, pipes and parts of roofs. By understanding the properties of metals, architects and builders can make the right decisions about which ones to use.

Other materials used to make our homes include polymers, ceramics and composites. As with metals, these materials are chosen carefully for their physical and chemical properties.

### Can we build homes without damaging the environment?

Most of the homes you see in the UK are made from bricks and mortar but not all of them. Earth materials such as straw or timber can also be used to build homes. In fact, around a third of the world's population lives in homes made from earth materials. Earth materials are becoming an increasingly attractive choice for building. This is because they have less negative impact on the environment than bricks and concrete. More recently, sustainable building materials such as 'hempcrete' have been developed. These materials are growing in popularity because homes built with them cost less to heat.

## Fuels used for cooking, heating and transport

### Why is oil so important?

We rely on crude oil for our society to function. Some of the products of crude oil are used as fuels to drive cars and aeroplanes. These include petrol, diesel and kerosene. Some of them are used as fuels in the home,

like methane and propane. These materials are called hydrocarbons – they contain only hydrogen and carbon.

### What happens when we burn fuels?

When we burn hydrocarbons in plenty of air, we get two products – carbon dioxide and water. This is because the hydrogen and carbon split up and react with oxygen in the air. The carbon dioxide produced by burning fossil fuels is a growing problem. The UK alone produces millions of tonnes of carbon dioxide every day. This and other greenhouse gases are increasing the likelihood of climate change. Also, the fossil fuels are non-renewable and they will run out. This will start happening in your lifetime.

## Generation and distribution of electricity

### How does crude oil help provide electricity to run my computer?

Hydrocarbons release lots of energy when they react with oxygen. This energy can be used to make electricity in power stations. Generally, this involves turning water into steam, which makes a turbine go round. The turbine is connected to a generator, which uses huge magnets to transfer the kinetic energy into electrical energy.

### What alternatives to crude oil are there?

Burning hydrocarbons to make electricity has two problems: the gases it produces and the fact that we're running out of fossil fuels. By using radioactive materials to turn water into steam we can reduce our use of hydrocarbons. This doesn't produce any carbon dioxide at all. However, the deadly radioactive waste will last for tens of thousands of years.

So how else can we make the electricity our society depends on? Wind, sun, tides, rivers and the Earth's crust can all be used to generate electricity. As they will not run out, we say they are renewable. However, each has its own advantages and disadvantages.

### How does electricity get to our homes?

Once electricity has been generated, it has to reach the consumer. High currents make wires hot and waste energy, so the current is reduced (stepped down) by devices called transformers. Thousands of miles of wire carry the electricity around the country. However, there are some concerns about whether this is safe.

As 44 per cent of the UK's $CO_2$ emissions come from households, we need to make homes more energy efficient. Energy use and sustainability are factors considered whenever new buildings are built. By choosing materials carefully, architects and construction companies can reduce their environmental impact.

Millions of pounds are spent on skiing and snowboarding holidays every year. New boards and skis are being invented all the time using space-age materials. The winter sports industry relies on new composite materials constantly being developed. It makes the sport more fun and ensures people will keep buying new equipment.

I live underneath electrical lines between pylons. Recently I've been reading in news stories that living near pylons could be dangerous. Apparently living here there is more risk of getting serious illnesses like leukaemia. Is this being properly researched? Are the government doing everything they can to make sure my family is safe?

Shell's 'Efficiency Improver' fuel was introduced in 2010 to help petrol last longer. It contains special additives that lubricate and clean engines. This means the engine wastes less of the petrol's energy overcoming friction. The less energy wasted, the further a car can travel on a tank of petrol.

# 7.1 Limestone as a building material

## Learning objectives

- How do we get limestone from the ground?
- Why is limestone such an important building material?

**Limestone** is one of the most important materials used in the building industry. It is used to make buildings and tiles. Even more is used to make cement which makes concrete. Concrete is the most widely used material in the building industry.

The main substance in limestone rock is **calcium carbonate**. Its chemical formula is $CaCO_3$. Most of it comes from the shells of ancient sea creatures. They died and sank to the bottom of the sea. The shells built up over hundreds of millions of years and eventually turned into **limestone rock**.

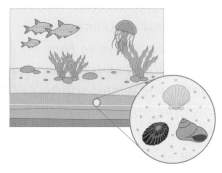

**Figure 1** Most limestone used to be the shelly parts of sea creatures

**a** What elements are present in calcium carbonate?

**Quarrying** companies extract more than 100 million tonnes of limestone from the ground in the UK every year. Explosives are used to blast limestone from the steep sides of the quarry.

**Figure 2** Limestone is quarried from the ground using explosives

**b** What happens at a quarry?

Construction engineers can use limestone directly as solid blocks. One problem with this is that limestone reacts with acid. Rain is usually slightly acidic, so it breaks down the surface of the limestone.

Materials engineers can solve this problem by turning it into other materials: **cement**, **concrete** and **mortar** are all made with limestone. Limestone is also used in making **glass**.

**Figure 3** Limestone is attacked by acid

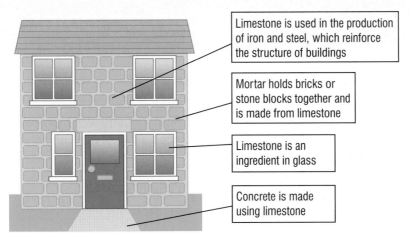

Limestone is used in the production of iron and steel, which reinforce the structure of buildings

Mortar holds bricks or stone blocks together and is made from limestone

Limestone is an ingredient in glass

Concrete is made using limestone

**Figure 4** A conventional house couldn't be built without limestone

- Cement is a starting point for making mortar and concrete.
- Mortar is the material that joins bricks together in a building.
- Concrete is used for making the structures of buildings. It is often made stronger by reinforcing it with steel.
- Glass is made from sand, with limestone and sodium carbonate added.

## Practical

### Are all limestones the same?

Limestone is extracted from quarries all over the world. Is each source of limestone the same? You can investigate this using limestone's reaction with acid.

You will need some limestone chips (from different places), hydrochloric acid, a conical flask and a balance.

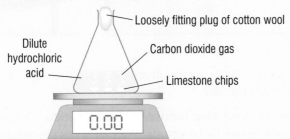

Loosely fitting plug of cotton wool

Dilute hydrochloric acid

Carbon dioxide gas

Limestone chips

0.00

Calcium carbonate reacts with acid to produce a salt, water and carbon dioxide. The more calcium carbonate there is in the limestone, the more carbon dioxide will be produced. This can escape through the top of the flask, making the mass decrease. You can compare the decrease in mass for different types of limestone.

### ∞ links

*For more information on Building materials see 7.2 Limestone as a starting point and 7.3 Products of limestone at work.*

## Summary questions

1  a  List **five** uses of limestone.
   b  Which of your answers to **a** can you see around you at the moment?

2  Write a flow chart showing how an ancient seashell can one day become part of a window.

3  Suggest which is more useful as a building material, iron or limestone? Refer to uses of each in your answer.

### Key points

- Limestone (containing calcium carbonate – $CaCO_3$) is blasted from the ground in quarries.

- Limestone is used to make buildings, concrete, cement, mortar and glass.

## 7.2 Limestone as a starting point

### Learning objectives

- What are quicklime and slaked lime?

- How is limestone converted into quicklime and slaked lime?

Lots of different technicians, engineers and scientists work directly with limestone. Their job is to make sure it is turned into useful products for construction.

First of all, limestone has to be extracted from a quarry with high explosives. Then, workers can crush it into smaller pieces. Most of the limestone goes to a cement works.

**a** What is most of the limestone from a quarry used for?

### Making quicklime for cement

It is my job as manager of this cement works to make sure everything runs smoothly. First of all, a materials manager organises our supply of limestone. Most of the limestone goes into our lime kiln. Production engineers operate the lime kiln. It heats the limestone up to around 1500 °C (about the same temperature as molten lava). This decomposes the limestone, turning it into a material called quicklime.

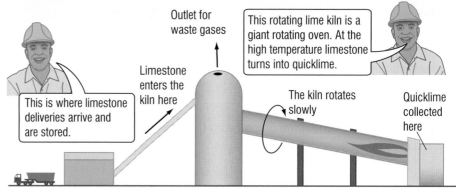

Outlet for waste gases

This rotating lime kiln is a giant rotating oven. At the high temperature limestone turns into quicklime.

Limestone enters the kiln here

This is where limestone deliveries arrive and are stored.

The kiln rotates slowly

Quicklime collected here

**Figure 1** At a cement works

The chemical name for quicklime is calcium oxide. Its formula is CaO. Quicklime is used to make cement.

The reaction in the lime kiln is called **thermal decomposition**. That's because heating is used to break down the limestone. We can show the reaction with this equation:

$$\text{calcium carbonate} \xrightarrow{\text{heat}} \text{calcium oxide} + \text{carbon dioxide}$$
$$CaCO_3 \longrightarrow CaO + CO_2$$

**b** How is calcium carbonate changed into calcium oxide?

**c** Name **two** materials quicklime is used to make.

### Making slaked lime from quicklime

Quicklime is useful as a starting material to make cement. By adding small amounts of water, it can be changed into another useful material, slaked lime.

Slaked lime can then be used to make lime mortar used when renovating old buildings. It is also used to treat soil that is too acidic for plants to grow in. This is because slaked lime is an alkali.

**d** What are **two** uses for slaked lime?

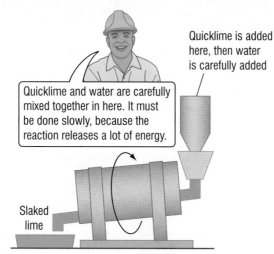

Quicklime is added here, then water is carefully added

Quicklime and water are carefully mixed together in here. It must be done slowly, because the reaction releases a lot of energy.

Slaked lime

**Figure 2** A lime slaking machine

The chemical name for slaked lime is calcium hydroxide. Its formula is $Ca(OH)_2$.

This equation shows how slaked lime can be made from quicklime:

calcium oxide + water $\longrightarrow$ calcium hydroxide

$CaO$ + $H_2O$ $\longrightarrow$ $Ca(OH)_2$

Carbon dioxide $CO_2$

Energy released

| Limestone $CaCO_3$ | | Quicklime $CaO$ | | Slaked lime $Ca(OH)_2$ |

Heat

Water $H_2O$

**Figure 3** This flow chart summarises the chemical changes when making quicklime and slaked lime from limestone

## Summary questions

1 Copy and complete using the words below:

*kiln   quicklime   carbon   energy   water*

Limestone can be changed into ............ by heating it in a ............ The limestone decomposes into quicklime and ............ dioxide. Quicklime can be changed into slaked lime by adding ............ This releases a lot of ............

2 Copy and complete the following table:

| Substance | Chemical name | Chemical formula |
|-----------|---------------|------------------|
| Limestone | | |
| Quicklime | | |
| Slaked lime | | |

3 Limestone and quicklime can both be ground down into a white powder. Think of a simple test you could use to tell the difference between them.

## Key points

- Quicklime is made by heating limestone to high temperatures.

- Slaked lime is made by adding water to quicklime. This releases energy.

- The chemical name for quicklime is calcium oxide and the chemical formula is $CaO$.

- The chemical name for slaked lime is calcium hydroxide and the chemical formula is $Ca(OH)_2$.

## 7.3 Products of limestone at work

### ??? Did you know ... ?

Touching cement with your bare hands can cause chemical burns. This is because of the quicklime it contains. Quicklime reacts with water giving off a lot of energy.

**Figure 2** Working with mortar

**Figure 3** Crystals of calcium hydroxide bind particles together when concrete and mortar set

### Making cement

Builders have used cement to bind buildings together for centuries. It was first used by the ancient Egyptians. They discovered that adding a little **gypsum** (**calcium sulfate**) to quicklime and water made a paste that was good at holding things together.

**Figure 1** The pyramids couldn't have been built without limestone

Today, cement is made by heating powdered limestone with **clay** and gypsum in rotating kilns.

Cement binds things together because the quicklime reacts with the water. Crystals of calcium hydroxide slowly grow out of each grain of cement. This forms a network which holds the whole structure together. Over time, the calcium hydroxide reacts with carbon dioxide in the air to form calcium carbonate.

**a** What are the three ingredients in cement?

### Using cement

You've probably seen cement mixers on building sites. Bricklayers use them to mix cement with sand and water to make a paste called **mortar**. The bricklayers spread the mortar onto bricks to hold them in place.

Concrete is similar to mortar, but also contains small stones called aggregate. Look around on your way home today. How many structures can you see that use mortar or concrete, and which is which?

**b** What would happen to mortar if too much water was used?

### Practical

**Which mix of mortar is strongest?**

Use different mixtures of sand and cement to make bars of mortar in a card mould.

Then test to find out which mix of mortar is the strongest.

## Making glass

Think of all the objects in your home made with **glass**. It's not just for windows. Glass is a versatile material that can be specially made to have a number of useful properties.

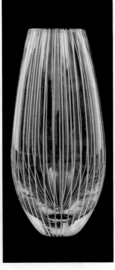

**Figure 4** Glass has many uses

**Figure 5** Thin pieces of glass can be brittle and fragile

Sand makes up over 60 per cent of glass. Limestone and sodium carbonate are added in smaller quantities. This mixture is heated up to 1500 °C. The raw materials react together to make glass. The molten glass becomes a solid as it cools down.

Glass is strong, but it isn't flexible. It is brittle. This can make it less useful for making objects that might get knocked about a lot. Glass can be made more flexible by using it to make a composite material such as fibreglass.

**c** What is the main ingredient in glass?

⚭ **links**

*For more information on composites see 7.6 Ceramics and composites in the home.*

AQA **Examiner's tip**

Don't get confused between concrete and cement. Remember – cement is used to make concrete.

### Summary questions

1 Copy and complete using the words below:

*hydroxide gypsum limestone clay water*

Cement is made by heating ............, ............ and ............. When ............ is added, crystals of calcium ............ grow through the mixture and hold it together.

2 What is the difference between mortar and concrete?

3 Complete this table

| Material | Ingredients | Uses |
|----------|-------------|------|
| Concrete | | |
| Mortar | | |
| Glass | | |

### Key points

● Limestone is heated with clay and gypsum to make cement.

● Sand can be added to cement and mixed with water to make mortar. Concrete is made by mixing sand, cement and stones together.

● Sand, limestone and calcium carbonate are heated together to make glass.

## 7.4 Metals for construction

### Learning objectives

- What are the properties of metals?
- How are metals used in construction?

AQA Examiner's tip

You need to be able to use data to explain why certain metals are chosen for certain jobs.

### Properties of metals

Most of the **elements** in the periodic table are metals. Scientists who work with metals are called **metallurgists**. Their role is to decide which metals are suitable for different jobs. They also combine metals with other metals (and some non-metals) to change their properties. These mixtures of metals, such as steel, are called alloys.

**Figure 1** Working with metals

The properties of metals make them ideal for use in the building industry. This table lists some of the properties most metals share:

| Property | Meaning |
|---|---|
| Hard | Metals are difficult to scratch |
| Strong | Metals don't break easily |
| Malleable | Metals can be beaten into different shapes |
| Ductile | Metals can be pulled out into wires |
| High melting point | Most metals don't melt easily |
| Good conductor of electricity | Electricity can travel through them easily |
| Good conductor of energy | Energy can travel through them easily |
| High density | They are heavy for their size because their atoms are packed closely together |

**a** Lightning rods are designed to channel lightning strikes harmlessly around a building. Which property makes copper a good choice for lightning rods?

### Metals in construction

Builders use metals alongside other materials. Look at the reinforced concrete being made in the photo opposite. Concrete is very strong in terms of compression. This means it can withstand large forces squashing it. However, pulling or twisting concrete can cause it to crack. Steel adds tensile strength to the structure, which resists pulling forces.

**b** If metal is so useful, why aren't buildings made entirely of metal?

**c** Why might it be useful to introduce a little flexibility into a building?

**Figure 2** This worker is making reinforced concrete

Here are some properties of metals found in a home. Concrete and wood have been added so you can compare them to metals.

| Material | Strength | Density | Malleability | Ductility | Electrical conductivity |
|---|---|---|---|---|---|
| Aluminium | ●●●●● | ●● | ●●●● | ●●● | ●●●● |
| Steel | ●●●●● | ●●●● | ●●●● | ●●●● | ●●●● |
| Copper | ●●●● | ●●●● | ●●●●● | ●●●●● | ●●●●● |
| Lead | ●●● | ●●●●● | ●●●●● | ●●● | ●●●● |
| Concrete | ●●● | ●● | | | |
| Wood | ●● | ● | | | |

**Key:**

| Very high | ●●●●● |
|---|---|
| Moderate | ●●● |
| None | |

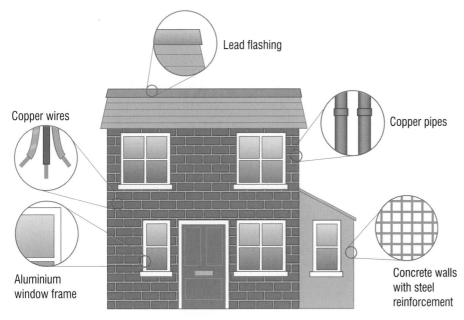

**Figure 3** Metals are used extensively in homes

Labels: Lead flashing; Copper wires; Copper pipes; Aluminium window frame; Concrete walls with steel reinforcement

## Practical

### Calculating density

Density is what makes the difference between a material being lightweight or very heavy. You can calculate the density of an object if you know its mass and its volume. You can measure mass easily using a balance. To measure volume, you need a displacement can. If you fill a displacement can up to the spout, water will spill out when you put another object in. You can collect the water in a measuring cylinder to find out the volume displaced by the object.

Once you have measured the volume, you can calculate an object's density using this formula:

$$\text{density} = \frac{\text{mass (g)}}{\text{volume (cm}^3)}$$

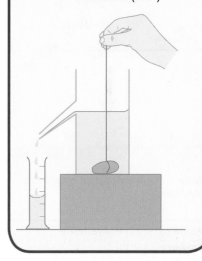

## Summary questions

1 Copper is malleable, ductile, a good conductor of electricity and does not react with water.
  **a** Which **two** of these properties make it good for making pipes?
  **b** Which **two** of these properties make it good for making wires?

2 Use the information in the table above to explain the different uses of metals in Figure 3.

3 Copper has the lowest chemical reactivity of the metals named on this page. Suggest why this is important for one of its uses in the home.

## Key points

● Metals all share similar properties. They are malleable, ductile, strong, hard, have high melting points and are good conductors.

● The properties of metals make them useful for many construction tasks.

# 7.5 Polymers in the home

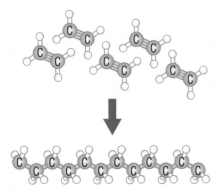

**Figure 1** Polymerisation of ethene into polyethene

As well as metals and limestone-based materials, **polymers** are used in the construction industry. They are also used in many consumer products found around the home. A wide range of materials allows architects, engineers and designers to be more creative in their projects.

Polymers are made of long-chain molecules. The small molecules that join together to make the polymers are called **monomers**. The process of turning monomers into polymers is called **polymerisation**. Most monomers in synthetic polymers come from crude oil.

Most polymers have these properties:

- low density
- flexibility
- low melting point
- waterproof
- chemically unreactive (resistant to corrosion)
- insulators of energy and electricity.

One of the simplest polymers is **polythene**. The scientific name for this polymer is **polyethene**. This is because it is made of many **ethene** molecules joined together in long chains.

**a** Where do the chemicals used to make polymers come from?

Figure 1 shows how part of a polyethene chain is made:

We can also show this in a chemical equation:

$$nC_2H_4 \longrightarrow (C_2H_4)_n \qquad \text{where } n = \text{a very large number}$$

ethene         polyethene
(monomer)     (polymer)

**b** What is polyethene made of?

This table summarises some of the properties and uses of common polymers.

| Name of polymer | Polyethene | Expanded polystyrene | Polypropene | PVC (polyvinylchloride) |
|---|---|---|---|---|
| Properties | Cheap, light and flexible | Soft, lightweight insulator | Hard and rigid | Cheap and flexible |
| Uses | Bags, wrappings, toys | Insulation and packaging | Car bumpers, high pressure pipes | Water pipes, shower curtains |

Polymers have replaced many more traditional materials since the 1900s. Polymers are often used for outdoor furniture instead of wood. This is because their low melting points make it easy to mould them into different shapes. Polymers used to make outdoor furniture are also waterproof, less dense than wood and resistant to weathering.

These properties make plastics useful for other uses outdoors, such as guttering, drains and water containers.

**Figure 2** Polymers often replace wood

**c** Why are polymers used for drains?

Plastic carrier bags have replaced paper bags because they're stronger. However, most supermarkets are now reducing the number of plastic bags they use. This is to reduce the environmental impact of making and disposing of them.

One of the largest uses of polymers in the home is for insulating electrical equipment. Polymers are excellent electrical insulators. They can also be moulded into any shape needed. This makes them perfect for insulating electrical appliances and wires.

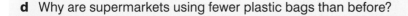

**AQA** *Examiner's tip*

You need to be able to explain why polymers are chosen for certain jobs.

**d** Why are supermarkets using fewer plastic bags than before?

**Figure 3** Polymers are important insulators

## Summary questions

**1** Copy and complete using the words below:

*monomers   chains   electronic   oil   guttering
electrical   outdoors   waterproof*

Polymers are long ............ made from smaller molecules called ............ Most monomers used to make plastics come from crude ............ Plastics are useful in ............ devices because they are ............ insulators and easy to mould. Plastics are useful ............ because they are ............ and resistant to weathering. Outdoor uses of plastics include ............, pipes and furniture.

**2** What properties make polymers a good choice for outdoor materials?

**3** Polymers are good insulators of energy as well as electrical insulators. Think of three uses of polymers in the home that rely on them being insulators of energy.

### Key points

- Polymers are made from the products of crude oil.
- Polymers are easy to mould and are excellent insulators.
- Polymers are also waterproof and have low densities.

# 7.6

# Ceramics and composites in the home

## Learning objectives

- What are ceramics and composites, and what are their properties?
- How do we use ceramics and composites around the home?

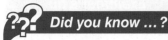

### Did you know … ?

Ceramic tiles are also used to heat-proof the outside of space shuttles.

Ceramics and composites are materials you may not have heard much about before. However, they have been used in homes for hundreds of years.

## Ceramics

**Ceramics** are materials like clay or china. This is probably the oldest branch of materials science. People have been making pottery for thousands of years.

Builders use ceramics in the bathroom or kitchen. Tiles, sinks, toilets, plates and mugs are all ceramics.

The properties of ceramics are:

- electrical and thermal insulators
- strong
- very hard
- resistant to most chemicals
- very high melting point
- **brittle** (they can shatter if struck sharply).

> **a** Which of the properties above make ceramics useful for making plates?
>
> **b** Which of the properties above makes ceramics less useful for making plates?

## Why use ceramics?

The most common ceramic objects you will see in buildings are bricks. Bricks are made of clay. This is an ideal material because of its high melting point and its strength.

> **c** What are the advantages of building houses with bricks instead of wood?

You will also see a lot of ceramics in the bathroom. Ceramics are ideal materials for sinks, tiles and toilet bowls. This is because they are hard-wearing, easy to clean and resistant to chemicals in cleaning products.

## Composites

**Composites** are one of the newest and most exciting developments in materials science. A composite is made when two or more different materials are put together to complement each others' properties. This is usually when a material has a really useful property but can't be used by itself.

They are interesting materials because they combine the best properties of different materials.

**Glass reinforced plastic** (often called **fibreglass**) is a composite material. It is made by painting a resin coating onto sheets of glass fibres. This results in a strong, yet flexible material. Fibreglass is used to make waterslides because it is stronger than plastic and just as light.

> **d** What properties of glass and polymers are important in fibreglass?

**Figure 1** Ceramics in action

**Figure 2** Tilers use the brittleness of ceramics to crack tiles

## Why use composites?

Because composites are made from a wide variety of materials, they have a wide range of properties. Their properties depend on the materials used, and also on how much of each one is used. The possibilities are endless.

When concrete is reinforced with steel it becomes a composite. Steel bars make buildings stronger because they add flexibility. They also slow down the growth of cracks. This is because the steel forms a barrier the crack has to move around.

**Figure 3** Water slides are made of a composite – glass reinforced plastic (fibreglass)

**Figure 4** Steel bars make it hard to break reinforced concrete

Thin metal wires can be placed in glass so it doesn't shatter if broken. This composite is called reinforced glass. Another way to reinforce glass is to add thin layers of plastic to its surface. This also holds the glass together if it gets broken. The plastic stops the glass breaking into sharp, dangerous shards.

Another composite found in the home is MDF (medium density fibreboard). MDF is often used to make 'flat-pack' furniture. It is useful because it is light, strong and easy to cut into shapes. It is made by binding individual fibres of wood together strongly with wax and resin. Because it does not have a 'grain' like wood, it does not chip or split easily.

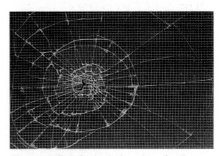
**Figure 5** Reinforced glass is safer than regular glass

**Figure 6** MDF is a composite used to make furniture

### _AQA Examiner's tip_

In the exam you will need to think about the properties of materials that are needed for the situation given in the question. This will help you to make your choices, descriptions or explanations relevant and gain more marks.

### Key points

- Ceramics are hard but brittle. They are good insulators. They are resistant to chemicals and have a high melting point.

- Composites are mixtures of more than one material. They often combine the best properties of each material in the mix.

### Summary questions

1. Which objects in your kitchen are made from ceramics?

2. Which properties of ceramics make them useful for bathroom tiling?

3. Ceramics are used to line kilns and furnaces. Why are they a good choice for this?

4. Make a table to summarise the information on this spread. Use the headings 'Material', 'Description', 'Properties', 'Uses'.

# 7.7 Building sustainable homes

## Learning objectives

- What are sustainable building materials and how do they compare to traditional materials?

- What are the features of straw bale, wood frame and cob houses?

There are more **sustainable** alternatives to modern construction materials. These sustainable materials are similar to more traditional building materials. They require less energy to make and put less carbon dioxide into the air. They also rely less on processed materials like metals and concrete.

## Wood versus brick

One way to tell if activities are sustainable (balancing our needs against care for the environment) is to compare their **carbon footprints**. This is a measure of how much carbon dioxide is put into the atmosphere by the activity. The diagram below compares some of the features of a brick house with a house made with more timber (wood).

> **a** What is a carbon footprint?

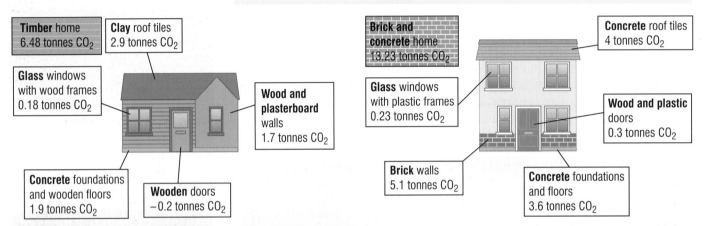

**Timber** home 6.48 tonnes $CO_2$

**Clay** roof tiles 2.9 tonnes $CO_2$

**Glass** windows with wood frames 0.18 tonnes $CO_2$

**Wood and plasterboard** walls 1.7 tonnes $CO_2$

**Concrete** foundations and wooden floors 1.9 tonnes $CO_2$

**Wooden** doors − 0.2 tonnes $CO_2$

**Brick and concrete** home 13.23 tonnes $CO_2$

**Concrete** roof tiles 4 tonnes $CO_2$

**Glass** windows with plastic frames 0.23 tonnes $CO_2$

**Wood and plastic** doors 0.3 tonnes $CO_2$

**Brick** walls 5.1 tonnes $CO_2$

**Concrete** foundations and floors 3.6 tonnes $CO_2$

**Figure 1** Carbon footprint of building a two-bedroom semi-detached house from wood or timber *Data adapted from Forestry Commission Scotland's ECCM report 196 (2006), Carbon benefits of timber in construction.*

## Activity

### Comparing wood and brick homes

Look at the two homes in Figure 1 above.

1 Draw up a table which compares the amount of $CO_2$ released during the production of each part of the two houses.

2 Using your table, make a bar chart of the data, then answer these questions.

3 Which part of each home has the biggest carbon footprint?

4 Which part of both homes has the biggest difference in carbon footprint between wood and brick?

> **b** What is the difference in carbon footprint between a brick house and a timber house?

Timber has such a small carbon footprint (see Figure 1) because trees absorb carbon dioxide during photosynthesis. If enough of a home is made from timber, the $CO_2$ locked up in the wood can balance the $CO_2$ released while building the home. A home in balance like this is called **carbon neutral**.

## Cob and hempcrete

As well as saving energy in construction, sustainable homes can use less energy to heat and cool. **Cob construction** is a good example of this. A cob building is made of a mixture of soil, clay and straw, with a thatched roof. As cob houses are usually built using local materials, less energy is needed to make them.

A modern alternative to cob is a substance called **hempcrete**. Hempcrete bricks are made from hemp (related to the marijuana plant) mixed with slaked lime. Like wood, hemp is carbon negative. A drawback with hempcrete is that it has only 20 per cent of the strength of concrete.

Because the walls of a cob house are thick, they absorb energy. This means the house is cooler during the day, saving money on fans or air conditioning. When the temperature goes down at night, the walls are still warm. This means less energy is needed to heat the house.

Controlling temperature like this is called **passive solar heating**. Hempcrete is not as good as cob at passive heating but is better than bricks.

## Straw bales and honeycomb clay blocks

Another sustainable way of building homes is using **straw bales**. The straw is tied together very tightly. Air trapped between the straws acts as an excellent insulator. As with cob houses, better insulation means much less energy is needed to heat the house.

A modern alternative to straw bales are **honeycomb clay blocks**. They insulate in a similar way, trapping energy in air spaces. One benefit of honeycomb clay blocks is they last longer than straw bales. However, they aren't carbon negative like straw.

> **c** Why do straw bale homes cost less money to heat?

Comparing different ways to build homes is complicated, because of the different factors involved. For instance, brick houses produce more $CO_2$ when built, but they last longer than wooden houses. The table below gives average figures for the energy usage and lifespan of different homes.

| Material | Lifespan (years) | Approximate carbon footprint of heating (tonnes of $CO_2$ per year) | Approximate carbon footprint of building (tonnes of $CO_2$) |
|---|---|---|---|
| Bricks | 150–500 | 2.5 | 13 |
| Timber | 150–200 | 2.1 | 7 |
| Straw bale | 90–150 | 0.6 | −5 |
| Clay honeycomb | 150–500 | 1.6 | 10 |
| Cob | 150–500 | 2.0 | 0 |
| Hempcrete | 90–150 | 2.2 | −10 |

**Figure 2** A modern cob home

**Figure 3** A straw bale house

## Summary questions

**1** Copy and complete using the words below:

*honeycomb footprint absorb clay hempcrete insulated lower longer*

Timber homes have a low carbon ............ because trees ............ carbon dioxide. Cob homes are made of local soil, ............ and straw. ............ blocks are similar to cob, but have a ............ carbon footprint. Straw bale homes are very well ............ because of air trapped in the bales of straw. Clay ............ blocks insulate in the same way as straw bales, but last ............

**2** Explain why cob homes can be described as being passively solar heated.

**3** Use the information in the table above to compare the following homes. Which would be better for the environment, and why?
**a** Brick versus straw bale    **b** Cob versus timber
**c** Clay honeycomb block versus hempcrete

### Key points

● Building conventional brick homes has a larger carbon footprint than using sustainable materials.

● Timber is a sustainable building material that 'locks up' carbon dioxide from when it was a living tree.

● Cob and straw bale houses use local materials and passive heating and cooling.

# Summary questions

**1** Describe three ways limestone is important in building houses.

**2** Copy and fill in the spaces in these equations showing the conversion of quicklime into slaked lime:

**Word equation:**

calcium oxide + ............ $\longrightarrow$ calcium hydroxide

**Symbol equation:**

CaO + ............ $\longrightarrow$ ............

**3** Which building material is made from:

  **a** gypsum, clay and limestone

  **b** sodium carbonate, limestone and sand

  **c** cement, sand, stones and water

  **d** cement, sand and water

**4** **a** What is the difference between flexibility and malleability?

  **b** Steel is an alloy made from iron.

    **i** How is steel used to reinforce concrete buildings?

    **ii** Why is steel not used to make aeroplanes?

  **c** Why is aluminium a better choice for window frames than iron?

**5** Suggest a use for the following polymers. Name properties of the polymer that make it a good choice for each use:

  **a** polyethene

  **b** PVC

  **c** polystyrene

  **d** polypropene

**6** The picture shows a kitchen. Suggest the best material to make each of the labelled items out of. Choose from the materials in the box.

> *ceramic   composite   metal   polymer*

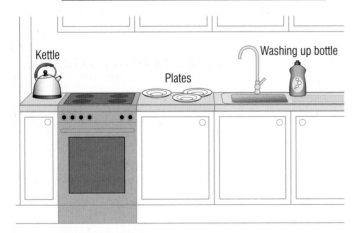

Kettle

Plates

Washing up bottle

**7** **a** The insides of furnaces are often lined with a ceramic material. Which property of ceramics makes them a good choice for this job?

**b** You work for a large corporation that supplies materials to other companies. These are the companies you supply materials to, and what they do with them:

| Company | What they produce |
|---|---|
| Alpha construction | Cement to sell to builders |
| Beta Bathrooms | Ornamental sinks |
| Gamma electrical | Electrical wires |
| Delta sports | Lightweight surfboards |
| Epsilon glass | Glass for windows |

These are the materials you have in your warehouse:

limestone  copper  clay  gypsum  fibreglass china  quicklime  sodium carbonate

  **i** Which material(s) would you send to each company? (Hint: some companies will need more than one material.)

  **ii** For Alpha Construction and Epsilon Glass, explain how they use the materials you are sending them.

  **iii** For Beta Bathrooms, Gamma Electrical and Delta Sports, explain why your choices of materials are best.

**8** **a** Straw bale and cob construction are traditional ways to build homes. Describe their properties and compare them to hempcrete and clay honeycomb blocks.

  **b** A carbon footprint is a measure of how much carbon dioxide an activity produces.

    **i** Why does a timber house have a smaller carbon footprint than a brick house?

    **ii** What other building materials have a small carbon footprint?

**9** A builder made notes about constructing a cob building from a website and now needs to sort them out. Read the notes and sort them in a table as shown.

| Advantage | Disadvantage | Not relevant |
|---|---|---|
|  |  |  |
|  |  |  |

> Straw is cheap, renewable and flammable.
>
> Straw walls need to be thick but it is easy to get hold of straw.
>
> Not strong in high winds, straw has been used as a building material for thousands of years.
>
> Straw bales are made from bundles of straw tied together with twine or wire and are a good insulator.

# AQA Examination-style questions

**1** Metals have many uses.
  **a** Explain why copper can be used for electrical wiring. Give two reasons. *(2)*
  **b** Explain why aluminium would be good for making aeroplanes. Give two reasons. *(2)*

**2** Fill in the table to show the chemical formulae for limestone and its products.

| Material | Chemical name | Chemical formula |
|---|---|---|
| Limestone | | |
| Quicklime | | |
| Slaked lime | | |

*(3)*

**3** Use words from the box to describe the composition of some materials.

> sodium carbonate   cement   clay
> gypsum   limestone   sand   water

  **a** Glass is made from ............. + ............. + ............. *(1)*
  **b** Mortar is made from ............. + ............. + ............. *(1)*
  **c** Cement is made from ............. + ............. + ............. *(1)*

**4** Plastic has many uses and replaces many traditional materials.
Suggest the traditional material for each of the modern uses of plastic and the advantage of using plastic in each of the situations.
  **a** Milk bottles *(2)*
  **b** Shopping bags *(2)*
  **c** Toys *(2)*

**5** Choose the correct words from the box to describe the uses of ceramics.

> are brittle   are resistant to chemicals
> are insulators   are malleable
> have a high melting point

  **a** Ceramics can be used for bathroom tiles because they ............. . *(1)*
  **b** Ceramics are used as furnace lining because they ............. . *(1)*
  **c** Ceramics can be used on power transmission lines because they ............. . *(1)*

**6** A shop selling hot drinks got a lot of complaints because the drinks it was selling got cold too quickly. The shop owners decided to do an experiment to find which material kept a drink hot the longest. The table shows their results.

| Cup material | Start temperature in °C | Temperature after 10 minutes in °C | Total drop in temperature in °C |
|---|---|---|---|
| China | 75 | 63 | |
| China with lid | 75 | | 5 |
| Plastic | 75 | | 21 |
| Plastic with lid | 75 | 58 | |
| Polystyrene | 75 | | 18 |
| Polystyrene with lid | 75 | 65 | |

  **a** Fill in the missing numbers in the table. *(2)*
  **b** Make and explain a conclusion based on their results. *(3)*
  **c** *In this question you will be assessed on using good English, organising information clearly and using specialist terms where appropriate.*

   Give a comparison of the advantages and disadvantages of making the cup out of china rather than polystyrene. *(6)*

**7** Glass-reinforced plastic (also known as fibreglass) can also be used in building houses. Fibreglass is made of a polymer reinforced by fine fibres made of glass.

It is strong and lightweight, it has a weather-resistant finish with a variety of surface textures and an unlimited colour range is available.

It can be used for roofing, door surrounds, canopies and chimneys.
  **a** Suggest why it is important to make sure no air is trapped between the glass fibres when the fibreglass is being made. *(1)*
  **b** Why might fibreglass be cheaper to install than other alternatives? *(1)*
  **c** Why might someone choose a wooden door surround? *(1)*

# 8.1 Everyday fuels

Learning objectives

- What fuels do we use and how do we use them?
- What are hydrocarbons and where do they come from?

**Figure 1** Natural gas provides energy for cooking

Our society relies on fuels in order to function. Homes in the UK use the equivalent of 45 million tonnes of **crude oil** every year. The total fuel use in the UK is around 150 million tonnes per year.

Without fossil fuels, we could not live the way we do today. This is because they are such a good source of energy. Burning fossil fuels releases lots of energy.

**a** Why do we burn fossil fuels?

## What fuels do we use?

The fuels you will come across every day include **natural gas** and **petrol**. However, there are many more in use in the UK. Here are a few:

| Fuel | Uses | Why do we use it? |
|------|------|-------------------|
| Natural gas (methane) | Domestic boilers, cookers | Can be compressed and stored, doesn't wear out pipes like liquids do |
| Petrol | Cars, taxis | Petrol engines are quieter and faster than diesel |
| Diesel | Cars, vans, buses, trucks | Cleaner and much more efficient than petrol |
| Heating oil | Domestic boilers | Can be stored in tanks where there isn't a permanent gas supply |
| Kerosene (paraffin) | Jet planes | Lower freezing point than diesel and burns more efficiently at high pressure than petrol |
| Coal | Heating homes, power stations | Easy to store, contains a lot of energy, doesn't easily catch fire |
| Propane | Gas barbeques, caravans | Easy to compress and store in tanks |

**b** Sort the fuels in the table into solids, liquids and gases.

## Fuels used at home

Most cookers in the home are powered either by gas or electricity. An advantage of using a gas cooker is that it is more responsive than an electric cooker. As gas cookers take less time to heat up, they are more efficient than electric ones.

Gas is also useful in heating our homes. This gas is either stored in a tank or supplied by mains pipes. The gas is burned in a boiler, which, in turn, heats water. The hot water is then either stored or used right away.

Not all homes are heated by gas, though. Remote places might not have access to a mains supply of gas. These homes often use heating oil as a fuel. This is because it is easy to store, usually in tanks that are buried underground.

Coal was once the main fuel used to heat homes. It was relatively cheap and easy to store, and contained a lot of energy. However, coal is dirty and awkward to use and homes needed to top it up regularly. Natural gas is cleaner and can be piped straight into our homes.

**c** Why are gas cookers more efficient than electric cookers?

## Fuels used for transport

**Petrol** and **diesel** are both used as fuels for transport. They are excellent sources of energy. They are denser than natural gas, so they are a more concentrated energy source. They are liquids, so they can flow, making them more useful than solid fuels because they can travel down pipes.

Inside a petrol or diesel engine, the fuel is mixed with air, and then ignited by a spark. This controlled explosion happens many times a second, transferring the energy to make the wheels go round.

**Kerosene** is used to fuel jet aircraft, but aircraft used to run on ordinary petrol. Over the years, jet engines have become more and more powerful. This means their fuel has to be more efficient. Kerosene burns more efficiently than petrol at high pressures.

**Figure 2** Inside a jet engine

## What are these fuels made from?

Most of the fuels we use for heating and transport are distilled from crude oil. This is a naturally occurring substance formed from the remains of plankton that died millions of years ago. The crude oil is now trapped in rock formations, usually deep underground.

These **reservoirs** of crude oil have to be drilled into and the oil pumped out of the porous rock formation. Only certain places on Earth have a good supply of crude oil. Billions of pounds are spent finding and exploiting new sources of crude oil. Crude oil is a mixture of **hydrocarbons**. These are chemicals containing carbon and hydrogen only.

> **d** What are hydrocarbons?

**Figure 3** Crude oil is the starting point for many fuels

### ∞ links

*For information on fractional distillation of crude oil look back at 2.4 Making products with materials from the Earth.*

### Summary questions

1 Copy and complete using the words below:

*diesel cooking burned kerosene energy carbon heating hydrocarbons*

When fuels are _____, they release a lot of _____. We use petrol, _____ and _____ for transportation. Natural gas and _____ oil are used for heating and _____ in the home. Fuels from crude oil contain _____, which are chemicals containing hydrogen and _____ only.

2 **a** Why do we use natural gas for domestic cookers?
  **b** Why do we use kerosene as jet fuel?
  **c** When is there an advantage in using heating oil instead of gas for domestic boilers?

3 Where did crude oil originally come from?

### Key points

● Fuels release a lot of energy when they burn in air.

● Petrol, diesel and kerosene are used for transport.

● Natural gas, heating oil, coal and propane are used for cooking and heating our homes and buildings.

● Hydrocarbons contain hydrogen and carbon only.

# 8.2 Burning fuels

## Learning objectives

- What new products are formed when fuels are burned?

- How can we describe burning as word or symbol equations?

- How can we show the complete combustion of a hydrocarbon as a balanced symbol equation? [H]

The reason we burn fuels is that they are such a good source of energy. Burning a fuel makes it release much of that energy. We use the energy to generate electricity, run machines and vehicles and warm our homes.

Burning a fuel in air makes it react with oxygen. Another word for this is **combustion**. When materials combust, new products are formed.

### Demonstration

#### Combustion reaction

When hydrocarbons burn, the carbon they contain reacts with oxygen to make carbon dioxide. The hydrogen they contain reacts with oxygen to make water. We can show this by setting up the experiment below.

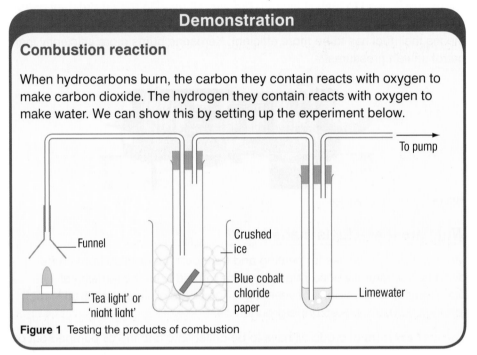

Figure 1 Testing the products of combustion

Cobalt chloride and limewater can both be used to test for chemicals:

- Limewater turns cloudy if carbon dioxide is bubbled through it.
- **Cobalt chloride paper** changes colour from blue to pink in the presence of water.

a What is combustion?

b What results would you expect to see if you tried the combustion reaction shown above?

Air hole completely closed; luminous flame

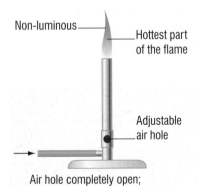

Air hole completely open; non-luminous flame, roaring flame

**Figure 2** Controlling combustion in a Bunsen burner

## Combustion of methane

You will have seen the combustion of methane before. It's exactly what happens on a gas stove or with a Bunsen burner.

You can see the importance of oxygen when you change the size of the air hole on a Bunsen burner. When the hole is closed, less oxygen can get in so combustion of the gas is not complete. The result is a sooty yellow flame. The soot is particles of carbon that have not combusted, and the yellow flame doesn't release much energy.

When the air hole on a Bunsen burner is open, more oxygen can get in. This means more complete combustion can happen. This gives you a very hot blue flame.

## Combustion equations

We can write out the combustion reaction as a general word equation:

hydrocarbon fuel   +   oxygen   ⟶   carbon dioxide + water

If the hydrocarbon is completely combusted, carbon dioxide and water are the only products that will be made.

methane   +   oxygen   ⟶   carbon dioxide + water

propane   +   oxygen   ⟶   carbon dioxide + water

Another way we can show this reaction is by using **'ball-and-stick' diagrams**. This type of diagram shows all the atoms and the bonds holding molecules together. It's another useful way to show how new products are made.

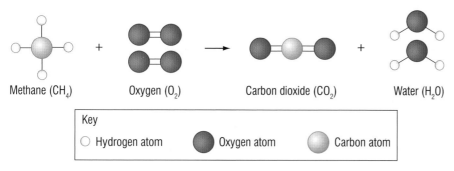

Methane ($CH_4$)        Oxygen ($O_2$)        Carbon dioxide ($CO_2$)        Water ($H_2O$)

| Key | | |
|---|---|---|
| ○ Hydrogen atom | ● Oxygen atom | ○ Carbon atom |

**Figure 3** The combustion of methane using ball-and-stick diagrams

## Higher

## Balanced symbol equations

As well as showing this as a word equation, we can use a symbol equation. This is a better way of showing how the chemicals have changed.

methane   +   oxygen   ⟶   carbon dioxide   +   water
$$CH_4 + 2O_2 \longrightarrow CO_2 + 2H_2O$$

propane   +   oxygen   ⟶   carbon dioxide   +   water
$$C_3H_8 + 5O_2 \longrightarrow 3CO_2 + 4H_2O$$

## Summary questions

1  Copy and complete using the words below:

*hydrocarbon   oxygen   water   carbon   burn*

When hydrocarbon fuels ............, they react with ............ from the air. We call this combustion. When a ............ completely combusts, it always produces ............ dioxide and ............ .

2  Butane and pentane are hydrocarbons like methane.
   **a**  Write a word equation for the complete combustion of butane.
   **b**  Write a word equation for the complete combustion of pentane.

3  Why do we not see the water produced when a Bunsen burner is lit?

4  Write a balanced symbol equation for the complete combustion of ethane, $C_2H_6$. **[H]**

### ⚭ links

*You can see how to balance combustion equations in 8.3 Differences between hydrocarbons.*

### Key points

- Hydrocarbons react with oxygen when they burn.

- Complete combustion of a hydrocarbon always produces carbon dioxide and water.

# 8.3 Differences between hydrocarbons (k)

## Learning objectives

- What is the structure of a hydrocarbon?
- How can we explain the patterns we see in the combustion of hydrocarbons?
- How can we balance combustion equations? [H]

Hydrocarbons are a very big family of chemicals. Luckily, they are also a very straightforward family. There are strict rules about the names they are given. Also, there are easy to remember patterns in their appearance and behaviour.

As you know, hydrocarbons are compounds made of hydrogen and carbon only. The carbon atoms are joined together in chains. The hydrogen atoms are joined to the outside of the chains, like this:

**Key**
○ Hydrogen atom
● Carbon atom

**Figure 1** This hydrocarbon would have the formula $C_{14}H_{30}$

## Chain length

Crude oil is a mixture of lots of carbon chains of different lengths.

**a** Draw the structure of a hydrocarbon with the formula $C_5H_{12}$.

Here are some of the shortest chain hydrocarbons. They are examples of **alkanes**:

| Name | Number of carbons in chain | Formula | Structure | Energy released by burning |
|---|---|---|---|---|
| Methane | 1 | $CH_4$ | | 802 kJ/mol |
| Ethane | 2 | $C_2H_6$ | | 1437 kJ/mol |
| Propane | 3 | $C_3H_8$ | | 2044 kJ/mol |
| Butane | 4 | $C_4H_{10}$ | | 2659 kJ/mol |

Look at the link between the number of carbon atoms and the number of hydrogen atoms. However many carbon atoms there are, there will be twice that plus two hydrogen atoms. This is true for all alkanes.

We can write this as a general formula: $C_nH_{2n+2}$

So for a hydrocarbon with 60 carbon atoms, there would be 122 hydrogen atoms (because $n = 60$, so $(2 \times 60) + 2 = 122$).

There are patterns in how chains of different lengths appear and behave. This comes in handy for separating them when crude oil is processed.

| Short chains | Long chains |
|---|---|
| Low boiling point | High boiling point |
| Catches fire easily | Doesn't catch fire easily |
| Thin and runny | Thick and viscous |

**b** Which is easier to ignite, $C_9H_{20}$ or $C_{20}H_{42}$?

There are also patterns in the combustion of hydrocarbons. The longer the chain, the more oxygen is needed. This means that larger hydrocarbons need a better air supply when they are burning. For example, a methane molecule needs two oxygen molecules during combustion, but a propane molecule needs five.

$$CH_4 + 2O_2 \longrightarrow CO_2 + 2H_2O$$
methane

$$C_3H_8 + 5O_2 \longrightarrow 3CO_2 + 4H_2O$$
propane

## Balancing combustion equations

*Higher*

You need to be able to work out how much oxygen is needed, using the formula of the hydrocarbon. You also need to work out how much carbon dioxide and water will be produced. Here are some tips:

- There must be the same number of carbon, oxygen and hydrogen atoms on each side of the arrow: i.e. in the propane example, there are 3 × C atoms, 8 × H atoms and 10 × O atoms on each side.
- Balance the carbons first: i.e. $C_3H_8$ must make $3CO_2$ (3 × C atoms on each side).
- Balance the hydrogens next: i.e. $C_3H_8$ must make $4H_2O$ (8 × H atoms on each side).
- Calculate how much $O_2$ is now needed to balance the O in the products: i.e. there are 10 × O atoms in $3CO_2$ plus $4H_2O$, so $5O_2$ are needed to balance the equation.
- If you end up with a 0.5 $O_2$, just double everything so you have only whole numbers

| | | | | | |
|---|---|---|---|---|---|
| *Let's try butane …* | $C_4H_{10}$ | + $O_2$ | $\longrightarrow$ | $CO_2$ | + $H_2O$ |
| Balance the carbon: | $C_4H_{10}$ | + $O_2$ | $\longrightarrow$ | $4CO_2$ | + $H_2O$ |
| Now balance the hydrogen: | $C_4H_{10}$ | + $O_2$ | $\longrightarrow$ | $4CO_2$ | + $5H_2O$ |
| Calculate the oxygen: | $C_4H_{10}$ | + $6.5O_2$ | $\longrightarrow$ | $4CO_2$ | + $5H_2O$ |
| Double to remove the 0.5: | $2C_4H_{10}$ | + $13O_2$ | $\longrightarrow$ | $8CO_2$ | + $10H_2O$ |

## AQA Examiner's tip

Look at the amount of carbon dioxide and water released and the amount of oxygen needed for the combustion of different alkanes. Make sure that you will be able to describe the patterns in the exam.

## Summary questions

1 What is the formula of an alkane with six carbon atoms?

2 Match up these formulae to the number of oxygen molecules they need to burn completely:

$C_{13}H_{28}$    $14 O_2$
$C_7H_{16}$    $20 O_2$
$C_9H_{20}$    $11 O_2$

3 Complete and balance this equation for the complete combustion of pentane. **[H]**

$$C_5H_{12} + O_2 \longrightarrow \text{----------} + \text{----------}$$

4 a Using the information on p128, draw a line graph of the number of carbon atoms in a hydrocarbon and the amount of energy obtained from burning it.

   b Use your graph to predict the amount of energy obtained from burning pentane (five carbons).

## ⚭ links

*For information on balancing equations look back at 2.9 Using equations.*

## Key points

- Alkane hydrocarbons have the general formula $C_nH_{2n+2}$
- Longer chains are darker, thicker and harder to boil or burn.
- The longer the chain, the more oxygen it needs when it combusts.

### 8.4    Problems with fossil fuels

Fossil fuels take millions of years to form. This means we can't replace the ones we are using now. We say they are **non-renewable**. This also means that there is only a limited amount of fossil fuel in the Earth's crust. Fossil fuel reserves will not last forever.

If we continue to use fossil fuels at our current rate and no new reserves are found, it is estimated that we will run out of oil and natural gas in the next 50 years.

### Oil is not just a fuel

An extra problem is that crude oil is not just used for fuel. It is also used for making many important chemicals, including plastics and medicines. As we run out of crude oil, we will need to find alternative ways to make these. Think of the plastic products you use every day – crude oil has a huge impact on your life.

**links**

*For information about using crude oil to make plastics look back at 7.5 Polymers in the home.*

As well as being important for making things, oil creates work for thousands of people in the UK. The petrochemical industry generates billions of pounds in the UK every year.

### Environmental effects of fossil fuels

Another problem with burning fuels is the pollution caused by the gases that are made. In Chapter 1 you saw that carbon dioxide traps energy inside our atmosphere. This is because it absorbs the energy the Earth radiates out into space.

On the one hand this is vital because it keeps the planet warm. On the other hand, too much greenhouse gas warms the Earth too much. This can cause climate change, resulting in more extreme weather systems all around the world.

> **a**   Why can greenhouse gases cause climate change and how does it affect us?

This table shows how much greenhouse gas the UK has produced between 1990 and 2009:

| Year | 1990 | 1995 | 2000 | 2005 | 2006 | 2007 | 2008 | 2009 |
|---|---|---|---|---|---|---|---|---|
| Equivalent amount of $CO_2$ (millions of tonnes) | 774 | 714 | 674 | 655 | 650 | 641 | 628 | 575 |

The **Kyoto Protocol** is an agreement between 37 countries (including the UK) to reduce greenhouse gas emissions. The UK has agreed to decrease its emissions to 92 per cent of 1990 levels by 2012.

Fossil fuels don't just produce carbon dioxide when they are burned. Other waste products from fuels can also be harmful to us and our environment.

- **Sulfur dioxide and nitrogen oxides** – produced by cars and power stations. Cause acid rain.
- **Soot** – tiny particles of carbon and unburnt hydrocarbons (also called particulates) made during incomplete combustion. Can cause breathing problems.
- **Carbon monoxide** – another product of incomplete combustion. Can cause brain damage or even death.

**Maths skills**

**Percentages of greenhouse gas levels**

Work out 92 per cent of the UK's 1990 greenhouse gas emissions. Is it a big decrease compared to current levels? Are we on track to meet it?

$$774 \times \frac{92}{100} = 712 \text{ million tonnes}$$

The UK beat this target in 1996!

**b** Which atmospheric pollutants are caused by transport?

Finally, fossil fuels can cause even greater problems when things go wrong. In 2010, the Deepwater Horizon, a floating oil drilling rig in the Gulf of Mexico, exploded. Eleven people died in the explosion, which also blew a hole in the sea floor.

Nearly 10 million litres of oil gushed from the hole per day. This caused an environmental disaster. Many habitats and countless thousands of animals were destroyed.

**Figure 1** The Deepwater Horizon disaster in 2010

**links**

*For information about the greenhouse effect look back at 1.7 Maintaining our atmosphere.*

## Summary questions

1 Copy and complete using the words below:

*nitrogen spills carbon dioxide acid fossil greenhouse*

All fuels produce ............ when they burn. Some also produce sulfur dioxide and ............ oxides. These gases add to problems like ............ rain and the ............ effect. Using ............ fuels can also result in environmental disasters like oil ............ .

2 **a** Use the table opposite of the UK's greenhouse gas emissions to draw a line graph.
   **b** Draw a dotted horizontal line at 92 per cent of 1990 levels to show the progress the UK is making.

3 Complete the following table using the information on these pages about crude oil.

| Social impacts (how does this affect me?) | Economical impacts (how does this affect jobs, business and money?) | Environmental impacts (how does this affect our environment?) |
|---|---|---|
|  |  |  |

### Key points

- Fossil fuels are running out. They are non-renewable.

- Burning fossil fuels produces carbon dioxide (the main greenhouse gas), which can affect our climate.

- Fossil fuels can harm the environment in other ways, such as acid rain or oil spills.

# 8.5    Generating electricity

## Learning objectives

● What are fossil fuels?

● How is electricity generated in fossil fuel power stations?

We rely on fossil fuels to provide most of our electricity in the UK. **Fossil fuels** formed over millions of years from the remains of animals and plants. Coal and gas are the main fuels used in the UK to produce electricity. Fossil fuels are useful because they release energy when they are burned. Deposits of coal, oil and gas are found worldwide, including the UK, although these supplies are limited.

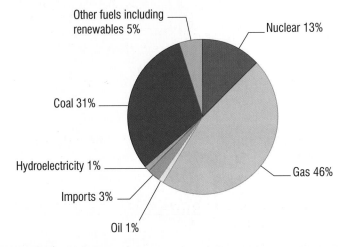

**Figure 1** Sources of fuel used for generating electricity in 2008 (based on information from *Digest of UK Energy Statistics 2009*)

If you move a coil of wire and a magnet near each other, an electric current is generated in the coil. Power stations spin a massive electromagnet inside a wire coil to generate electricity. This is the **generator**.

A **turbine** is linked to the generator. The turbine is like a giant fan or windmill with many blades. When the turbine spins, the generator spins too. This generates electricity. The main difference between power stations is how we spin the turbine.

**Figure 2** This turbine spins when jets of steam or gas are forced through it

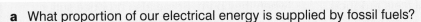

   **a**   What proportion of our electrical energy is supplied by fossil fuels?

In fossil fuel power stations, we burn the fuel. This heats water. The water changes into steam, which blasts past the blades of a turbine, making it spin. All fossil fuel power stations have a stack (chimney). **Flue gases** from the burning fuel are released through the stack. Some steam is cooled then recycled and reheated in the pipes. Large **cooling towers** cool the rest of the steam before it is released into the atmosphere.

New gas-fired power stations are more efficient than coal-fired power stations. The gas is burned and the hot exhaust gases force the turbines to spin. The waste gases are still hot. They are used to heat water, changing it to steam, which also spins turbines.

   **b**   What is the job of (i) the generator, (ii) the turbine, (iii) the stack, (iv) the cooling tower?

   **c**   Why are gas-fired power stations more efficient than coal-fired power stations?

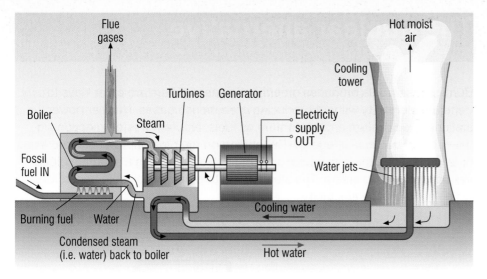

**Figure 3** How electricity is generated in a fossil fuel power station

## Activity

### Advantages and disadvantages of fossil fuels

Use your textbooks and other resources to find out advantages and disadvantages of burning fossil fuels.

Some advantages and disadvantages will be more significant and have a greater impact than others.

Use your ideas and information to place these advantages and disadvantages of burning fossil fuels in a rank order. Include reasons for your rank order. Some of these problems can be overcome, so look out for information on cleaner power stations and other ideas. This information may help you change your order.

Prepare a presentation either using PowerPoint, as a leaflet or as a poster. Present this information to your class, and compare your rankings and reasons with others in your class.

### Did you know ... ?

Half of the UK's 20 coal-fired power stations are on the coast because of the large quantities of water required to generate electricity.

## Summary questions

1 Copy and complete using the words below:

*boiler   furnace   generator   turbine*

In fossil fuel power stations, the fuel is burned in the ............. to heat water in the ............. This produces steam which spins a ............. The turbine turns the ............., producing electricity.

2 Write down one similarity and one difference between a gas-fired power station and a coal-fired power station.

3 Suggest why the amount of electricity generated using gas is increasing each year and the amount generated using coal is decreasing.

### Key points

● Fossil fuels release energy when they burn.

● Several stages take place in a fossil fuel power station before the electricity is generated.

## 8.6

# The nuclear alternative

Burning fossil fuels produces greenhouse gases. There are other ways to generate electricity without producing greenhouse gases. Nuclear power stations use **uranium** and **plutonium** as fuels. **Nuclear fission** reactions in the nuclear fuel release energy. Depending on the type of power station, a gas or liquid is piped through a heat exchanger. Here, water in a separate piped system changes to steam. The steam turns turbines, which spin generators, which generate the electricity.

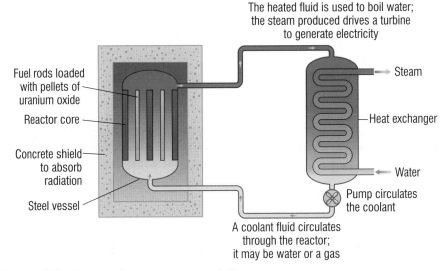

**Figure 1** What happens in a nuclear power station

### Activity

**Nuclear power**

Prepare a leaflet explaining how a nuclear power station generates electricity.

---

**a** Name **two** nuclear fuels.

## Advantages of nuclear power

There are many advantages to using nuclear power. It is a reliable form of energy, and does not release greenhouse gases. There are currently plenty of supplies of nuclear fuels. Using nuclear power conserves our fossil fuel supplies.

Nuclear power is very safe. The only accident to cause deaths occurred nearly 25 years ago in Ukraine. Several hundred people died as a direct result of the accident. There was also a noticeable increase in thyroid cancers for several years afterwards. There have been close calls in USA and some minor accidents in the UK.

This compares with thousands of deaths per year while mining for coal in China alone, and over 170,000 deaths caused when the Banqiao hydroelectric dam in China collapsed in 1975.

**??? Did you know ... ?**

France generates 78 per cent of its electricity by nuclear power, producing about a third of the carbon dioxide per person per year compared to America.

**b** State **two** advantages of using nuclear power.

## Disadvantages of nuclear power

There are some disadvantages to nuclear power. Many safety precautions are built into the design of nuclear power stations, so the cost of producing electricity is high.

It is very expensive to decommission nuclear power plants safely at the end of their working lives. Several of the UK's nuclear power plants will need to be decommissioned in the next ten years.

If there were to be a nuclear accident the economic and environmental damage would be enormous. It would take tens, if not hundreds, of years to recover from.

There are some radioactive emissions from nuclear power stations. However, emissions from coal-fired power stations are 100 times more radioactive.

The main drawback of nuclear reactors is the highly radioactive waste produced from the used fuel rods.

> **c** State **two** disadvantages of using nuclear power.

## Radioactive waste

Low-level radioactive waste is not dangerous to handle and is disposed of in landfill sites.

Medium-level waste is solidified in concrete or bitumen, then buried underground in special storage areas.

Highly radioactive waste has to be stored safely for years while the radioactivity dies away. Some can be recycled, but the rest is vitrified (changed into glass). After 50 years, the waste is about 1000 times less radioactive, but still needs to be safely stored for hundreds of years. Relatively small amounts of this highly radioactive waste are produced.

**Figure 2** A nuclear fuel recycling plant at Sellafield

**Figure 3** Nuclear waste is changed into glass (vitrified) and stored to stop radioactivity seeping into the surroundings

## Summary questions

1 Copy and complete using the words below:
   *rods generators exchanger turbines fission*

   In nuclear power stations, nuclear ............ reactions cause fuel ............ to heat up. A heat ............ uses this energy to change water, in pipes, into steam. The steam spins ............, which turn ............ to produce electricity.

2 Write down one similarity and one difference between a nuclear power station and a fossil fuel power station.

3 Explain one benefit and one drawback of nuclear power on the environment.

## Key points

- Nuclear power generates electricity using nuclear fission reactions.

- There are advantages and disadvantages in using nuclear power.

- Nuclear power is safe and does not produce greenhouse gases. It produces radioactive waste.

# 8.7 Renewable energy resources

We are surrounded by natural resources that can be used to generate electricity. As we have become more dependent on electricity, we need resources that will not run out – these are **renewable energy** resources.

Electricity is generated when a generator connected to a turbine spins. When wind, water or steam flows over blades connected to a turbine, electricity is generated. The energy source is free for most renewable energy sources. However, the structures built to generate the electricity can be very expensive for the amount of electricity generated.

**a** What is the difference between a renewable energy source and a non-renewable energy resource?

## Burning biomass

**Biomass** power stations generate electricity when wood, poultry litter and straw are burned. The energy released changes water to steam which spins the turbines, generating electricity. One disadvantage is that carbon dioxide (a greenhouse gas) is emitted from the power stations. However, if plant matter is burned, the plants have absorbed carbon dioxide as they grew. The energy source is reliable and a single power station can generate large amounts of electricity. The cost of a power station is similar to the cost of a fossil-fuel power station.

## Wind farms

**Wind farms** can be found near the coast, on hills or offshore. They use a free source of energy that is non-polluting. Large blades on wind turbines spin when it is windy generating electricity. To improve efficiency, turbines can be up to 100 m tall.

However, these turbines cannot produce energy if it is calm or very stormy. The noise they cause can be very disturbing and they are often very visible against the natural landscape.

## Hydroelectric power

**Hydroelectric** schemes use dams to trap water in reservoirs in mountainous areas. The water is released when needed through pipes in the dam wall. Turbines in the pipes spin as the water travels through them.

Hydroelectric dams produce electricity very quickly when needed to cope with surges in demand. In quiet periods, some hydroelectric schemes pump the water back up into the reservoir so it is ready for the next time it is needed. Hydroelectric schemes are reliable and non-polluting, but can only be sited in mountainous areas.

The biggest disadvantage with hydroelectric power is that, to trap water, local habitats have to be flooded. This often means people have to move from surrounding towns and villages.

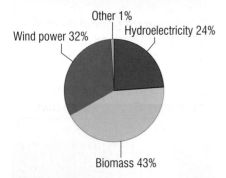

Other 1%
Wind power 32%
Hydroelectricity 24%
Biomass 43%

**Figure 1** How our renewable energy was provided in 2008

**Figure 2** The UK is planning to build more wind farms offshore

**Figure 3** Valleys in mountainous regions are flooded when hydroelectric schemes are developed

## Tidal power

**Tidal** schemes use large barriers across the mouths of rivers. These are called tidal barrages. They force seawater through pipes when the tide goes in or out. Turbines in these pipes spin when the water is flowing, and electricity can be generated for about 40 per cent of the time. The schemes are reliable and can generate large amounts of electricity.

However, they are expensive to build, and there are only a limited number of places in the world where the schemes would work. The river flow and habitat around the estuary are damaged by flooding.

## Geothermal power

**Geothermal** schemes are most successful in volcanic regions. Shafts are sunk very deep into the Earth's surface. Cold water is pumped down these deep shafts and there is enough heat transferred from hot rocks to change the water to steam. It escapes up another shaft to spin turbines in a power station. In the UK, some geothermal energy is used to heat water rather than produce steam.

## Wave power

At the moment, there are no large-scale electricity supplies powered by waves, although the technology is developing. Some sites in Scotland have been identified for the first commercial-scale developments. These should be completed by 2020.

Although wave power is non-polluting, it is unpredictable and it has proved hard to harness the energy from the sea.

## Solar power

**Solar cells** generate small amounts of electricity directly from sunlight. The cells are expensive for the amount of electricity generated. However, solar cells are useful in devices like calculators, which only use a small amount of electricity. They are also useful on satellites or rural bus stops, which are remote from other sources of energy. The cells can be useful in very sunny climates but have limited uses in the UK.

b Which types of renewable energy can generate electricity at any time when required?

c Which forms of renewable energy can generate a lot of electricity in the UK?

**Figure 4** The world's first tidal scheme was in France. Only a few schemes have been built worldwide.

### Did you know … ?

There has been a geothermal energy scheme in the UK since the 1980s. In Southampton, enough energy has been retained underground for useful amounts to be extracted and used to heat houses, shops, hotels and leisure centres. Electricity from the scheme will be used in Southampton port.

**Figure 5** The Pelamis generator will start producing commercial wave energy in Scotland by 2020

### Key points

- Renewable energy sources will not run out.
- The main renewable energy sources are biomass, wind, hydroelectricity, tidal, geothermal, solar and wave power.
- There are advantages and disadvantages in these schemes.
- Renewable energy sources use free resources and do not produce greenhouse gases. They have an impact on the environment.

### Summary questions

1 Copy and complete using the words below:

*biomass geothermal hydroelectric solar tidal waves wind*

a These types of renewable energy can only be used in certain places: ............ and .............

b These forms of renewable energy can only be used at certain times: ............ and .............

c These types of energy depend on the weather: ............ and .............

2 Write down one advantage and one disadvantage for each renewable energy source.

## 8.8 The National Grid

### Learning objectives

- How does electricity get to our homes?
- Does the National Grid affect our health?
- Does the National Grid damage the environment?

The **National Grid** is used to deliver electricity throughout the country. **Pylons** are large metal towers that support the cables which carry electricity from **power stations** to homes and factories. Power stations supply the electricity at several thousand volts. Then **step-up transformers** increase this to even higher voltages.

Increasing the voltage lowers the current. A lower current means that less energy is wasted by the cables getting hot although the same power is supplied. This makes the National Grid more efficient with less energy transfer from overhead cables. **Sub-stations** near homes contain **step-down transformers** to reduce the voltage to a safer level, 230 V.

> **a** Why is electricity transferred in the National Grid at high voltages?

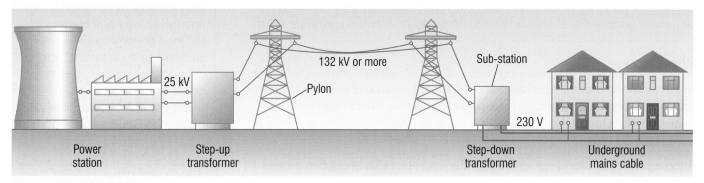

**Figure 1** The National Grid

### Is there a health risk?

Cancers in children may have many causes, such as a mother having X-rays or smoking during early pregnancy. Two separate studies have given us a cause for concern, although there is no proven link.

Children living within 100 m of high voltage cables were compared with other children. The scientists found that the children living near the cables were nearly twice as likely to suffer from leukaemia as children living further away (leukaemia is a cancer of the blood).

These studies compared thousands of children with cancer born over more than 25 years in two countries. In some cases, the cancers occurred in groups of people living near each other (a cluster). This made experts wonder if the cause was a factory, or pollution in the surrounding area.

> **b** Explain why many people trust these studies.

### Electric and magnetic fields

Electric and magnetic fields are places where electric or magnetic forces can be detected. These fields surround high-voltage cables as well as electrical equipment. They are weaker further from the equipment, and when the current or voltages are lower. Experts do not believe the high voltages from the cables are causing the cancers because:

- Electric fields from the high-voltage cables cannot pass through buildings.
- Magnetic fields due to high-voltage cables are about 100 times weaker than the Earth's magnetic field.

**Figure 2** People living close to high-voltage cables may worry about the effects of electric and magnetic fields

- Experiments using electric and magnetic fields at similar levels did not cause changes to living cells.

**c** What are electric and magnetic fields?

## Activity

### Cable health risk

Write a leaflet for **one** of the situations below:

**1** From an electricity company explaining to local residents why high voltage cables are not a risk to health.

**2** From a pressure group of parents of children with leukaemia explaining why they think that high-voltage cables are a risk to health.

## What are the environmental problems with maintaining the National Grid?

Pylons supporting high-voltage cables can be 50 m high and so stand out against the natural landscape. Some are painted to try to merge with the sky. The area surrounding the cables is kept free from trees and other structures. This is so the cables can be maintained safely and are less likely to be damaged.

In some conditions, power cables hum or crackle, which can disturb people living nearby. The high-voltage cables and pylons can be damaged by wind, ice and lightning. Damaged cables can cause power cuts as well as fire or electrocution.

**d** List **three** environmental effects caused by the National Grid.

**Figure 3** Trees near power cables are cut back to prevent damage and power cuts during storms

## Activity

### Underground or overhead?

Discuss the advantages and disadvantages of burying electric cables underground. Decide what research is needed, and how you will present your findings. Keep a record of each website you use and decide how reliable it is. In small groups, pool your information and present it to the class.

## Summary questions

**1** Copy and complete using the words below:

*increase   lines   pylons   reduce*

The National Grid is a network of ............ and high voltage power ............. Step-up transformers ............ the voltage before it is carried by the power lines. Step-down transformers ............ the voltage before it is supplied to homes.

**2** Explain why the evidence does not prove that living near high voltage cables causes cancer.

### Key points

- Electricity is supplied through the National Grid.

- The National Grid includes power stations, sub-stations, pylons and cables.

- Electromagnetic fields surround high voltage cables.

- Some health risks, and damage to the environment, are increased near high voltage cables.

## Summary questions

**1** Fuels are chosen because they have the right properties for a particular task.

   **a** Explain why many cars use petrol instead of diesel, even though diesel is more efficient.

   **b** Explain why kerosene is used as a fuel in jet planes.

**2** When natural gas (methane) burns, what products are made? How could you prove it?

**3** Dodecane is a hydrocarbon with 12 carbon atoms.

   **a** What is its chemical formula?

   **b** Would dodecane have a higher or lower boiling point than hexane (six carbon atoms)?

   **c** Explain the reasoning of your answer to part b.

**4** Pentane is a hydrocarbon with the formula $C_5H_{12}$.

   **a** Write a word equation for the combustion of pentane.

   **b** Write a balanced symbol equation showing its complete combustion in oxygen.    **[H]**

**5** Our society relies heavily on fossil fuels for energy.

   **a** Fossil fuels produce pollutants when they burn. List five of these pollutants and explain the harm they cause.

   **b** How have the UK's greenhouse gas emissions changed over the past 20 years?

**6** Explain what the role is for these parts of a coal-fired power station.

   **a** furnace

   **b** boiler

   **c** turbine

   **d** generator

**7** Write down two differences and two similarities between a coal-fired power station and a nuclear power station.

**8** **a** Explain how a wind turbine generates electricity.

   **b** Write down one advantage and one disadvantage of using wind turbines to generate electricity.

**9** Put these parts of the National Grid in the order that electricity passes through them from a power station to the home.

- step-down transformer
- step-up transformer
- power cable
- power station
- house
- sub-station.

**10** Match the diagram of the energy source A–D to its name.

| Energy source | Diagram |
|---|---|
| Geothermal |  |
| Hydroelectric |  |
| Wave power | A |
| Wind power | B |

# AQA Examination-style questions

**1** There are many different ways to produce energy by burning fuels.

    **a** Give examples of **two** renewable fuels. *(1)*

    **b** Name **two** fossil fuels. *(1)*

    **c** Name a fuel that can be used without burning it. *(1)*

    **d** Complete a balanced symbol equation for the combustion of pentane ($C_5H_{12}$). **[H]** *(4)*

**2** Match each type of energy source up with the correct disadvantage.

| Energy source | Disadvantage |
|---|---|
| Fossil fuel | Waste stays dangerous for a long time |
| Nuclear fuel | Can be a problem distributing the electricity |
| Tidal | Stops fish swimming up river to spawn |
| Wave power | Produces greenhouse gases |

*(3)*

**3** The diagram shows the generation of electricity in a power station.

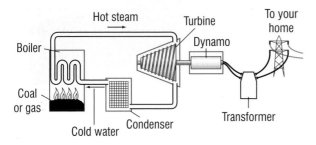

    **a** Use the diagram to help you describe how to generate electricity in a power station. *(4)*

    **b** The national grid uses cables to distribute electricity. The electricity that passes through these cables is at a high voltage. Why? *(3)*

**4** A caravan owner did an experiment to find out which make of camping gas was the best value for money.

| Make of gas | Cost in £ | Time it lasts in days |
|---|---|---|
| Campagas | 28 | 20 |
| Gas-on-the-go | 27 | 18 |
| Porta-gas | 30 | 24 |
| Travelgas | 19 | 15 |

    **a** Calculate the cost per day for Campagas. *(2)*

    **b** Explain which gas was the best value for money. *(2)*

    **c** Choose **one** thing that the caravan owner could do to make the experiment more accurate.

| Only do the experiment at the weekends | Only buy one kind of gas | Repeat the experiment | Time how long the gas lasts in seconds |
|---|---|---|---|

*(1)*

**5** Read this article from a local newspaper, then answer the questions.

## Uproar over new pylons

Electricity pylons are used to distribute electricity all over the world yet in one Yorkshire town there have been mass demonstrations against a new line of them across the dales. Dr David Winter, spokesperson for Greenenergy said 'if everyone used alternative ways of generating their own electricity, then we would not have this problem'. Dr Winter himself has a solar panel and a wind turbine on the roof of his house. Another alternative to using pylons is underground cables. These are high voltage cables that are buried deep underground. However, underground cables sometimes get really hot which can cause damage to them. This is addressed by pumping water or oil along the cable to cool it.

    **a** What are **two** advantages of having underground cables? *(2)*

    **b** Explain why using alternative ways of generating electricity would be much better for the environment. *(2)*

    **c** *In this question you will be assessed on using good English, organising information clearly and using specialist terms where appropriate.*

    Suggest why people like Dr Winter have chosen to install solar panels and wind turbines on the roofs of their houses but other people do not. *(6)*

# My property

In Unit 2, Theme 3 you will work in the following contexts, covered in Chapter 9:

## The cost of running appliances in the home

### How can we find out how much our electricity bill is?

Many people are sent an electricity bill every three months. The bill uses readings from our electricity meter to calculate how much we owe. We can also work out which equipment adds most to the bill.

### How quickly do we use electricity?

We are often told to cut down on the electricity we use to reduce greenhouse gas emissions. Different equipment doing the same job can use very different amounts of electricity. How can we compare the electricity used by different appliances so we can make the correct choices?

### How can we choose the most efficient equipment?

Making clever choices in the shop saves pounds on running costs. All equipment transfers some of the electrical energy it uses into wasted forms. Buying energy-efficient equipment means we waste less of the electricity that we buy.

## Why are electrical appliances so useful?

Electricity is useful because it can can flow through wires to different places. Appliances at home transfer electrical energy to many different forms of energy. Find out how we can show these energy transfers using diagrams and how these can be used to compare different devices.

## Electromagnetic waves in the home

### How do we use electromagnetic waves in our lives?

Electromagnetic waves transfer energy at the speed of light. We use different types of waves for communicating, cooking and medical uses.

### Are electromagnetic waves harmful?

Different electromagnetic waves affect cells in our body in different ways. Used properly they are useful, in other situations they can be harmful. How can we use different electromagnetic waves while limiting any harm from them?

*Part of my job is helping customers choose the most suitable equipment for their needs. It's often the most efficient equipment too as this will save money on their electricity bills. We can use efficiency ratings to save customers' money.*

*My job as an electronics engineer means that I design and install communications networks using electromagnetic waves daily. Our networks use fibre optic cables and phone lines as well as wireless connections. We may set up temporary systems to cover events such as the Olympic Games, or permanent systems linking households and businesses.*

# 9.1 Energy

**Figure 1** This hairdryer transfers electrical energy into different forms

When you turn on the hairdryer, it is supplied with **electrical energy**, which transfers into different forms of energy. The types of energy from the hairdryer include energy heating the element and the air around it, **sound energy** and **kinetic energy** (movement). The total energy given out by the hairdryer matches the amount supplied. This **energy transfer** can be described using this equation:

**electrical energy** $\longrightarrow$ **energy for heating + sound energy + kinetic energy**

**a** Write down the energy equation for a light bulb. It transfers electrical energy into energy heating the filament and its surroundings and light energy.

In all cases, some of the final forms of energy are ones we want, but there are also unwanted transfers too. No energy is destroyed during any transfer. However, some is transferred to the surroundings and becomes less useful.

The amount of energy an object has indicates how much work it can do. We measure energy in **joules (J)**. To give you an idea, one joule is given to an apple when you lift it by about 1 m.

Larger amounts of energy are measured in **kilojoules (kJ)**. One kilojoule is 1000 J. An adult going up one floor in a lift gains about 2000 J, or 2 kJ.

**b** Change (i) 5000 J into kJ, (ii) 0.3 kJ into J.

**c** What happens to the total amount of energy during an energy transfer?

## Sankey diagrams

Sankey diagrams show these energy transfers in detail. They include an arrow for each of the different forms of energy. The width of each part of the Sankey diagram indicates the proportion of that type of energy. The Sankey diagram for a light bulb shown in Figure 2 shows that for every 100 J of electrical energy, 90 J transfers to heat and 10 J transfers to light.

Before you draw a Sankey diagram, you need to know the types of input and output energy. You also need to know the amount of each type of energy.

When you draw the Sankey diagram:

● the wide part of the arrow on the left represents the input energy
● the arrow splits to match the number of different output forms of energy
● the width of each arrow is proportional to the amount of each type of energy
● the unwanted forms of energy (usually heating an appliance and its surroundings and sound) are shown at the bottom.

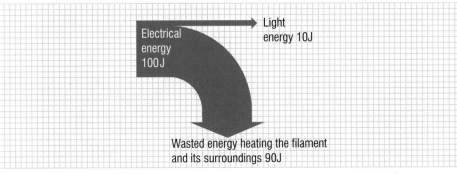

**Figure 2** The Sankey diagram shows details of the energy transfers in a light bulb

**d** What do you notice about the amount of input energy and the amount of output energy for the light bulb?

**e** Draw a Sankey diagram for an energy-efficient light bulb. It transfers 100 J of electrical energy into 70 J of heat and 30 J of light.

⊂⊃ **links**
*For more information about the efficiency of various household appliances see 9.4 Efficiency.*

**Figure 3** A coal-fired power station wastes more energy than it transfers as useful electrical energy

**f** Describe the energy transfers shown in Figure 3.

## Summary questions

1 Write down three things that a Sankey diagram tells us.

2 Draw a Sankey diagram for these energy transfers.
   **a** A radio transfers 250 J of electrical energy into 210 J of thermal energy and 40 J of sound energy.
   **b** A hairdryer transfers 100 J of electrical energy into 60 J of thermal energy, 35 J of kinetic energy and 5 J of sound energy.

### Key points

● Energy is measured in J or kJ.
● Sankey diagrams show the energy transfers taking place.

# 9.2 Electrical power

## Learning objectives

- What is power?
- What units is power measured in?
- How do we calculate power

**Figure 1** These bulbs have different power ratings, and use electricity at different rates

The equipment in our homes transfers energy at different rates. An electric heater transfers energy more quickly than a lamp. We say the electric heater is more powerful than the lamp. **Power** measures how quickly energy is transferred. It is measured in **watts** (W) or kilowatts (kW), where 1 kW = 1000 W. If the power of a device is 1 W, then it transfers one joule of energy per second.

Some equipment like a tumble dryer is very powerful, but only used for short periods of time. Equipment like freezers are less powerful, but are left on all the time. The energy transferred by the freezer in a year can be higher than the energy transferred by the tumble dryer. This means you could save more money than you might think on electricity bills if you choose a freezer with a low energy rating.

### Maths skills

**Measuring power**

power (W) = energy (J) ÷ time (s)

**Worked example**

A lamp transfers 1200 J in 20 seconds. What is its power?

Its power is 1200 ÷ 20 = 60 W.

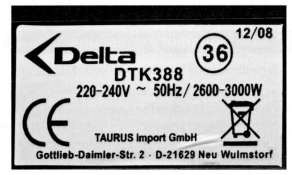

**Figure 2** This information tells you the power of this device

a What is the power of a heater that transfers 6 kJ in 6 seconds?

b How much energy does it transfer in 1 minute (60 seconds)?

 **Maths skills**

## Calculating electrical power

The power of electrical equipment can be calculated using:

**power (W) = potential difference (V) × current (A)**

For example, if a computer mouse charger is rated at 3.3 V and 0.3 A, what is its power?

Its power is 3.3 × 0.3 = 0.99 W

Electrical equipment usually has information about its energy usage printed on its base. The information is also printed on bulbs. It states the power rating, voltage and/or current used by the equipment when it is turned on.

The **potential difference** is often called the voltage. In the UK, our electricity is supplied at 230 V. In Europe and America, electricity is supplied at a different voltage so the equipment settings may need to be changed if an appliance is used abroad.

c Why doesn't equipment always use the same amount of power in America compared with the UK?

d When a kettle is plugged into the mains supply (230 V), its current is 10 A. What is the power of this kettle?

## Practical

### Calculating the power

Use a joulemeter to measure how much energy is transferred in one minute by low-voltage bulbs of different powers. Calculate the power of each bulb using your results and compare this with the theoretical amount. Repeat this with a motor or immersion heater.

### Did you know ...?

Electrical cookers use more power than most electrical devices in the home. Often they use their own circuit containing a 30 A fuse.

## Summary questions

1 Complete these sentences:
   a Power measures how quickly ............ is transferred.
   b A kilowatt is one ............ watts.
   c Electricity is supplied at 230 ............ in the UK.

2 Work out the power in each of these examples.
   a A radio transfers 250 J in 2 minutes.
   b A hairdryer uses a current of 9 A when it is plugged into the mains supply in the UK.

3 How much energy is transferred when a 100 W bulb is left on for 1 hour?

### Key points

- Power is the rate at which energy is transferred.
- Power is measured in W or kW.
- Power is calculated using: power = energy ÷ time, or power = voltage × current.

# 9.3 Buying electricity

## Learning objectives

- What is the unit of electricity?
- How do we calculate the Units of electricity used?
- How do we calculate the cost of electricity used?
- What are the implications of running home appliances?

All homes have an electricity meter to measure the amount of electricity used. Readings are taken or estimated four times a year. It is inconvenient to charge customers for electricity in joules. One joule is a tiny amount of energy. We are billed in Units instead. 1 Unit = 1 kilowatt-hour (kWh). It is the energy used if something with the power of 1 kW (1000 W) is switched on for 1 hour, or if a 100 W light is on for 10 hours.

**Units (kWh) = power (kW) × time (h)**

**a** How many Units are used if a 2 kW electric fire is on for 5 hours?

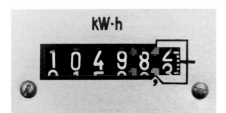

**Figure 1** This electricity meter shows the Units used since it was installed

The extra number of Units used since the last bill is used to calculate the bill. The number of Units used is the difference between this reading and the last reading.

Different companies charge different amounts per Unit used. The cost of the electricity is calculated using:

**cost of electricity = Units used × price per Unit**
**(pence) (kWh) (pence per kWh)**

The total bill usually includes a quarterly charge. This has to be paid whether you use electricity or not.

### AQA Examiner's tip

Remember to include the correct units with your answer.

**b** An electricity meter reads 34578. Three months ago it read 33235. How many Units have been used?

**c** If each Unit costs 14p, what is the cost of electricity used?

**Electricity Statement**

Electricity Readings

| Meter Serial no. | Read Date | *Read Type* | Read | Last Read | Units Used |
|---|---|---|---|---|---|
| | | Removal | 28619 | 28170 | 449 |
| | | Smart | 1749 | 0 | 1749 |
| **Total units** | | | | | **2198 kWh** |

Electricity Charges 05 Jan 2010 - 30 Jun 2010

| | | | | |
|---|---|---|---|---|
| Electricity supply standing charge | 177 days | 22.0p per day | £ | 38.94 |
| Electricity total unit charge | 2198.0 kWh | 8.085p per kWh | £ | 177.71 |
| **Total supply charges** | | | **£** | **216.65** |
| VAT @5.00% | | | £ | 10.83 |
| **Total cost of electricity** | | | **£** | **227.48** |

**Figure 2** This electricity bill gives information on electricity usage and costs

We can all make changes to reduce the size of our electricity bills. There are three main ways this can happen:

- We can use equipment more efficiently. This includes:
  - not leaving computers and TVs on standby
  - not leaving phone chargers on all night
  - only boiling the amount of water you need in a kettle.

- We can buy equipment that wastes less electricity. This includes:
  - using energy-efficient bulbs instead of filament bulbs
  - using equipment with a better efficiency rating, e.g. washing machines and freezers.
- We can use equipment less. This includes:
  - turning lights off when we leave the room
  - hanging clothes out to dry instead of using a tumble dryer.

Some savings make a lot of sense. Freezers and fridges are not powerful but they are on all the time. This means inefficient models can use quite a lot of electricity over a year. Electric heaters and tumble dryers are powerful so they add Units to the electricity bill quite quickly. Reducing the time that they are in use makes sense.

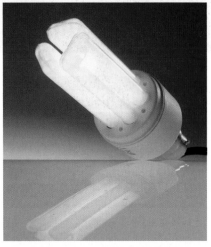

**Figure 3** Use energy-efficient bulbs to reduce the electricity used compared with filament bulbs

> **d** Write down **one** way to reduce your electricity bill for each of the three main methods. Use different examples to the ones in the text.
>
> **e** If you only had one energy-efficient bulb to use in your home, where would you use it? Explain.

Electricity is a clean energy source at the point of use. However, pollution is caused by power stations where electricity is generated. If all households used less electricity, the demand for electricity would fall, generating less pollution. However, the impact of a single household on the amount of electricity generated is too small to measure. Concentrate on the benefits of saving money at home instead!

**Figure 4** Reducing our electricity bills helps to reduce demand for electricity

**f** What is the main benefit to each household of reducing electricity usage?

## Summary questions

1 Explain why we buy electricity in kWh instead of J.

2 **a** How much electricity did Joel use? He had a 100 W light on for 4 hours, watched TV (500 W) for 2 hours and used his computer (750 W) for 2 hours.

  **b** Each Unit costs 15p. How much did Joel add to the electricity bill?

3 Write a paragraph explaining several ways that electricity bills can be reduced.

## Key points

- Electricity is sold in Units of kWh
- Units are calculated using power (kW) × time (h).
- The electricity bill is based on the Units used and price per Unit.

# 9.4

# Efficiency 🄺

## Learning objectives

- What does efficiency mean?
- How is it measured?
- Can I use appliances more efficiently?

**Figure 1** These lamps produce the same amount of light but use different amounts of electricity

### ?? Did you know … ?

Electrical devices can be 90–95 per cent efficient. However, the electricity is produced in power stations which are only about 30–60 per cent efficient. Many inefficient devices transfer unwanted energy, heating their surroundings, or as sound energy. Unwanted energy transfers can take place in the plugs and cables as well as in the equipment itself.

One of the easiest ways for many households to reduce their electricity bill is to swap over to using energy-efficient bulbs. These do not heat up as much as traditional bulbs. This means less electricity is used to provide the same light output.

Traditional bulbs only change about 5 per cent of the electrical energy transferred to them into light energy. The remaining energy is transferred to the surroundings, warming them up. We say they are 5 per cent **efficient**. This is one reason why 100 W bulbs for household use can no longer be imported into the UK, or supplied to shops. In contrast, energy-efficient bulbs are 25 per cent efficient.

About 20 per cent of our electricity bill is for lighting so we could save 10–15 per cent on electricity bills by using energy-efficient bulbs. Although they are more expensive initially, they last longer and have cheaper running costs.

**a** What percentage of energy does an energy-efficient bulb waste?

 **Maths skills**

### Calculating efficiency

Efficiency is calculated using either of these equations:

$$\text{efficiency} = \frac{\text{useful energy out}}{\text{total energy in}}$$

or

$$\text{efficiency} = \frac{\text{useful power out}}{\text{total power in}}$$

For example, a kettle provides 190 J of useful energy for every 200 J of input electrical energy. Its efficiency is:

190 J ÷ 200 J

= 0.95.

Efficiency does not have any units but it can be shown as a percentage. This is calculated as:

efficiency = useful energy out × 100% ÷ total energy in

The kettle's efficiency will be:

190 J × 100% ÷ 200 J

= 95%

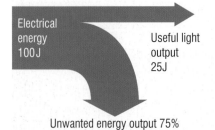

**Figure 2** A Sankey diagram for an energy-efficient bulb

Electrical energy 100 J

Useful light output 25 J

Unwanted energy output 75%

## Sankey diagrams again

You can use information from a Sankey diagram to calculate efficiency. It shows the input energy and all forms of output energy. You should only include the useful forms of output energy in the calculation.

Efficiency can never be more than 1.0, or 100%. This is because you can never get more energy out from a device than you put into it. Check your calculation carefully if your answer is greater than 1.0 or 100%.

**b** Draw a Sankey diagram for a traditional bulb that transfers 100 J electricity into 5 J light energy and 95 J energy heating the surroundings.

**c** What is the efficiency of a dishwasher if it uses 500 J of electrical energy for every 400 J of useful output energy?

**⬯⬯ links**

*For information on how to draw and interpret Sankey diagrams look back at 9.1 Energy.*

## Energy ratings

If you use inefficient electrical equipment, you will spend more on your electricity bills. Energy ratings are used to compare goods like fridges, washing machines, cookers, light bulbs and dishwashers. These can waste large amounts of electricity as they may be left on for long periods of time, or be very powerful.

Customers can use the efficiency rating to choose equipment with lower running costs for the same performance. An A-rated machine is very efficient, using less than 55 per cent of the energy of an E-rated machine.

The electrical equipment you choose makes a big difference to your electricity bill. A household that uses energy-efficient bulbs will reduce its total electricity bill by about 10–15 per cent. A household only using A-rated equipment could reduce its bills by another 10–15 per cent compared with one using only C-rated equipment. This means that about 25 per cent of an electricity bill could be saved just by using more efficient equipment.

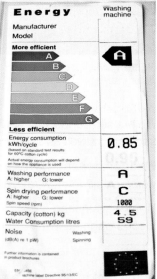

**Figure 3** The label tells the customer how efficient the washing machine is

**??? Did you know … ?**

The EU Energy Label is a compulsory notice applied to all white goods sold in the European Union. It allows people buying the goods to compare the efficiency and energy consumption of different models.

**d** Why is it useful to be able to compare the efficiency of different equipment before you buy one?

### Summary questions

1 Complete these sentences
   **a** Efficiency can never be more than ............ per cent.
   **b** Efficiency equals ............ energy output ÷ total energy input.

2 Explain what is wrong with this statement: Equipment with an efficiency rating of A always uses less electricity than equipment with a C rating.

3 The cost of a traditional light bulb is 50p, and it will last about 1 year. The cost of an energy-efficient bulb is £3, and it will last about 4 years. Explain why it is still worth buying the energy-efficient bulb.

### Key points

● Efficiency measures how much energy is usefully transferred.

● Efficiency can never be more than 100 per cent.

● Efficiency is calculated as useful energy output ÷ total energy input.

● You can reduce your electricity bills by using efficient equipment.

## 9.5  How fast do waves travel?

### Learning objectives

- What do waves transfer from place to place?
- How do we describe waves?
- How do we calculate the speed of waves?

**Figure 1** When a stone falls into a pond, waves transfer energy as they travel across the water's surface

**AQA** *Examiner's tip*

Remember that all the waves in the electromagnetic spectrum travel at the same speed. They just have different wavelengths and, therefore, frequencies.

We are surrounded by **waves** in nature. These include water waves, sound and light as well as seismic waves.

**Electromagnetic waves** are waves that include radio, microwave, infrared, visible, ultraviolet, X-rays and gamma waves.

Many waves need a material to travel through, such as a solid, liquid or gas. Electromagnetic waves can also travel through a vacuum.

All waves have features in common:

- They are a regularly repeated **disturbance** that passes through the material.
- They all transfer energy from one place to another, but the material doesn't travel with the energy.

**a** Write down **two** things that all waves have in common.

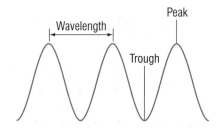

**Figure 2** A transverse wave

Transverse waves cause disturbances at right angles to the direction in which the energy travels. Water waves are one example of transverse waves. The surface of the water moves up and down but the energy travels along the surface to the edge of the pond.

The wavelength is the distance between one peak and the next, or between one trough and the next. It is measured in metres.

The frequency is the number of waves produced each second, or the number of waves passing a certain point each second. Frequency is measured in hertz. One hertz is a frequency of one wave per second.

**b** The frequency of one wave is 12 Hz. How many waves are produced each second?

All waves travel at a certain speed. This depends on how long each wave is (its wavelength) and how many waves are produced each second (its frequency).

 *Maths skills*

## Wave calculations

The velocity of a wave is calculated using:

**velocity (m/s) = wavelength (m) × frequency (Hz)**

What is the velocity of a sound wave? Its frequency is 1000 Hz and its wavelength is 0.34 m.

The velocity of the sound wave = 1000 × 0.34
= 340 m/s

All electromagnetic waves travel at 300 million m/s. This means that:

the frequency of any electromagnetic wave is 300 million ÷ wavelength.

the wavelength of any electromagnetic wave is 300 million ÷ frequency.

For example, calculate the frequency of a radio wave. Its wavelength is 1000 m.

The frequency = 300 million ÷ 1000
= 0.3 million Hz

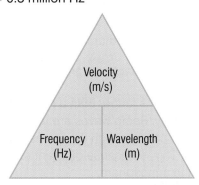

**Figure 3** The wave speed triangle

c   What is the speed of a water wave? Its wavelength is 1.5 m and its frequency is 0.5 Hz.

d   What is the wavelength of a microwave? It is a member of the electromagnetic spectrum, and its frequency is 10 000 million Hz.

 *Did you know ... ?*

Nothing can travel faster than the speed of light in a vacuum. The mass of objects, however tiny, increases so much at speeds near this that they can't accelerate any more.

## Summary questions

1   Complete these sentences choosing the correct word:
   a   **Water waves/microwaves** are an example of electromagnetic waves.
   b   Water waves are an example of **transverse/longitudinal** waves.
   c   Wave speed equals frequency **multiplied/divided** by wavelength.

2   Which of these waves is travelling faster? A wave with a frequency of 1000 Hz and a wavelength of 0.02 m, or a wave which travels 25 m in 2 s?

3   Sketch and label a transverse wave with a wavelength of 2 m.

## Key points

● All waves transfer energy without transferring matter.

● Waves are regular disturbances passing through matter.

● Wave speed = wavelength × frequency.

## 9.6

# Electromagnetic waves

### Learning objectives

- What is the electromagnetic spectrum?
- What are the uses of the electromagnetic spectrum?
- Why do we use the electromagnetic spectrum?

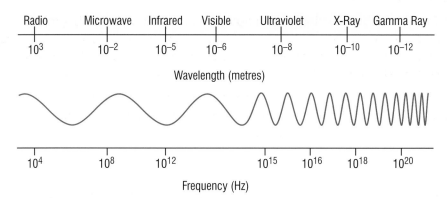

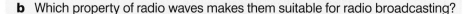

**Figure 1** The electromagnetic spectrum

**Figure 2** Mobile phone networks use microwaves that travel between mobile phone masts placed in a network throughout the country

**Electromagnetic waves** are waves that travel through space at the speed of light, 300 million m/s. These waves form a continuous spectrum, but are placed in seven main groups. The properties and uses of the different groups of waves depend on their wavelengths and frequencies.

Properties that depend on the type of electromagnetic wave include:

- the amount of energy they carry,
- whether they are absorbed, transmitted or reflected by different materials.

**a** Write down **three** properties of electromagnetic waves.

Most **radio waves** have wavelengths longer than a metre. Their frequency is so low they carry very little energy. Radio waves are used to transmit radio programmes because they can pass through the atmosphere. Their range depends on their wavelength. This is because different radio wavelengths reflect off different layers in the atmosphere.

**b** Which property of radio waves makes them suitable for radio broadcasting?

**Microwaves** have wavelengths of between about a centimetre and a metre long. They carry more energy than radio waves. Since microwaves are not reflected or absorbed by the atmosphere they can communicate with satellites. This is how we receive satellite TV broadcasts.

Mobile phone networks use microwaves to send signals between mobile phone masts throughout the country. Wi-fi systems also use microwaves to link computers wirelessly within buildings.

Food is cooked in microwave ovens because microwaves are absorbed by water, fat and sugar molecules, heating the food.

**Infrared** wavelengths range from about a centimetre to millionths of a metre. TV remote controls and some wireless communications use infrared waves. Hotter objects emit more infrared radiation than their surroundings, so infrared cameras help find suspects after dark and casualties in a fire.

**c** What property of microwaves makes them suitable for satellite TV broadcasting?

### AQA Examiner's tip

A common mistake is to think that infrared or radio waves are used to communicate between mobile phones. Infrared can be used for file sharing over short distances but if you text or call someone up, it is microwaves that transfer the energy.

**d** What property of infrared makes it suitable for finding suspects at night?

**Visible light** is detected by our eyes so we can see. Its wavelength is about a millionth of a metre.

Visible light is used in **fibre optic cables** for communications. Fibre optic cables are very fine strands of glass, bundled together in cables. Light passing through these cables repeatedly reflects off the inside surface of the glass. This is called **total internal reflection**. Very little energy is wasted in the cable.

These cables are used in **endoscopes** to look inside the body during keyhole surgery. Internet and telephone connections, as well as cable TV, are transmitted using fibre optic cables.

**e** Explain what is meant by total internal reflection.

**Ultraviolet radiation** reaches us from the Sun. It is also produced by sun beds. Its wavelength is about ten billionths of a metre. It is absorbed by our skin, which responds by tanning. Cells on the surface of tanned skin absorb UV radiation effectively, preventing it from reaching the cells deeper in our skin.

**X-rays** are very energetic. Their wavelength is about a tenth of a billionth of a metre. An X-ray image is a shadow picture. The light patches show bones which X-rays could not pass through. The darker patches show places where X-rays could pass.

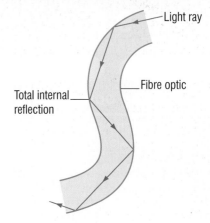

**Figure 3** Optic fibres carry many messages simultaneously

Labels: Light ray, Fibre optic, Total internal reflection

**Figure 4** Radiotherapy treats cancer by killing cells using gamma rays

The wavelength of a gamma ray is about a million millionth of a metre. **Gamma rays** and X-rays have enough energy to kill cancer cells. Ways of using gamma rays in cancer treatment include implants near the tumour, or beams of gamma radiation directed at the tumour from outside the body. Gamma radiation is used to sterilise medical equipment as it kills bacteria.

## Did you know … ?

Technology develops so fast that there are new developments in wireless technology every month. Devices such as the iPad use electromagnetic waves.

## Key points

- The energy of electromagnetic waves increases with frequency.
- Radio waves are used for TV and radio broadcasts.
- Microwaves are used for mobile phone networks, satellite TV and cooking food.
- Infrared waves are used in TV remote controls.
- Visible light is used in fibre optic cables.
- Ultraviolet is used in sunbeds.

## Summary questions

1 Complete these sentences by choosing the correct word:
   **a** Microwaves have a **longer/shorter** wavelength than radio waves.
   **b** Infrared waves carry **more/less** energy than microwaves.
   **c** Radio waves have a **higher/lower** frequency than microwaves.
   **d** Optic fibres use **visible light/ultraviolet** radiation.
   **e** Skin cancer is caused by **infrared radiation/ultraviolet** radiation.

2 Describe **one** use for each of infrared, radio waves, microwaves, visible light and ultraviolet light. Write down the property that makes each type of wave suitable for this use.

# 9.7

# Dangers of radiation ⓚ

## Learning objectives

- Can electromagnetic waves be dangerous?
- How do we use X-rays and gamma rays?
- Why are X-rays and gamma rays not used in the home?

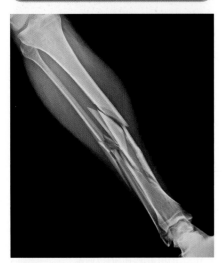

**Figure 1** X-rays reveal damage to bones

**Figure 2** X-rays and gamma rays can only be used by authorised workers in controlled areas

## Shorter wavelengths

As the wavelength of electromagnetic waves gets shorter, the energy carried by the waves increases. The waves are absorbed by our cells. Higher energy waves have a bigger effect on our bodies and can cause harm.

**a** Why can shorter electromagnetic waves cause harm?

X-rays and gamma rays pass through a patient's body. They are used to diagnose and treat different medical problems without operating on a patient.

## X-rays

X-rays are mainly used in a hospital or dentist's surgery because too much exposure can be harmful, and their use must be controlled. If an X-ray is needed, the parts of the body that are not being X-rayed are shielded.

**b** What property of X-rays means they are not suitable to use at home?

## Gamma rays

Cancer cells can grow rapidly and stop nearby organs from working properly. Gamma rays and X-rays have enough energy to kill cells so doctors use these to target the cancer cells. It is important that nearby healthy cells are not damaged. So treatment is designed to shield these cells or expose them to a lower dose of radiation.

**c** How are healthy cells protected during cancer treatment?

When X-rays and gamma rays are absorbed, they can damage or kill cells. Damaged cells may mutate and become cancerous. This means the use of X-rays and gamma rays must be controlled. To reduce harm, doctors limit how often a person is treated. They aim to use as low a dose as possible. This reduces the chance of long-term damage.

Different patients and different cells are more vulnerable. Unborn babies and children are vulnerable because they are growing so their cells are dividing rapidly. Damage in children and adults is harmful if a vital organ like the heart, kidney or liver stops working properly. When a patient is treated, lead screens shield areas that are not being treated.

Workers in X-ray and radiotherapy departments must also be protected. They are frequently exposed to small doses of radiation.

**d** Why can X-rays and gamma rays be harmful?

## UV radiation

Ultraviolet radiation is less energetic than X-rays but still causes harm. A tan is the way skin cells try to protect themselves from exposure to UV radiation. Doctors are worried because there are more skin cancer sufferers nowadays. Skin cancer can take years to develop. Doctors think over-exposure to the Sun during overseas holidays and in sunbeds caused damage that is now turning cancerous in some people.

UV from the Sun is very intense in hotter countries. People can protect themselves by covering up, staying inside during the early afternoon and using sun cream. Some regular exposure to sunlight is important, however. It helps us to produce vitamin D in our bodies, which is essential for the formation of bones and teeth.

**e** Why are some people more likely to suffer from skin cancer after holidays in hot countries?

Figure 3 Microwaves from mobile phones are unlikely to be harmful

## Microwaves

Microwaves are not very energetic but may possibly still cause harm. Harm may be caused from a high dose over a short period of time. Microwave ovens cook food because molecules in cells absorb microwaves and heat up. Damaged microwave ovens may let some microwaves seep out around the door seals. This is unlikely to be harmful as the dose is extremely low.

Some people are worried that low doses of microwaves over many years from mobile phones may cause harm. So far, studies suggest the risk, if it exists, is tiny. The radiation emitted by mobile phones is too low to cause cancers directly. Mobile phone use has increased dramatically but the incidence of brain tumours has not. However, cancer takes years to develop and the way the studies have been carried out has not convinced everyone.

**f** Why may some people be unaware that ultraviolet waves are harming them?

### Summary questions

**1** Complete these sentences choosing the correct word:

*X-rays gamma radiation ultraviolet waves microwaves*

............ can be used to treat cancer. ............ and ............ are too dangerous to be used in the home. ............ are given out by mobile phones.

**2** Write down three pieces of advice that a person who works in an X-ray department could follow to reduce the risk of harm to patients from X-rays.

**3** Write down three pieces of advice that a mobile phone user should follow to reduce any possible risk of harm from microwaves.

**??? Did you know ... ?**

Mobile phones are linked to an increased risk of death. This is because car drivers making or receiving calls are more likely to have accidents because they are distracted.

### Activity

**Are mobile phones dangerous?**

Investigate the evidence that mobile phones may cause tumours or cancers. Present your data as a poster suitable to be displayed in school.

### Key points

- Microwaves and ultraviolet radiation can be harmful.

- Hospitals use X-rays to produce shadow pictures of bones and gamma rays are used to treat cancer.

- X-rays and gamma rays are harmful in large doses so their use is controlled.

# Summary questions

**1** Draw a Sankey diagram to show this energy transfer: a lamp transfers 10 J of electrical energy into 2 J of light energy and 8 J of heating the lamp and its surroundings energy.

**2** The power of a kettle is 2000 W. It is used for 15 minutes each day. The power of a fridge is 25 W. It is on for 24 hours. Which appliance uses more energy?

**3** Calculate the missing values in the table.

| Power in watts | Current in amps | Voltage in volts |
|---|---|---|
| 45 | 0.9 | |
| | 1.3 | 14 |
| 62 | 8 | |

**4** These readings were taken from an electricity meter.

21 June

| 2 | 6 | 3 | 2 | 4 | 2 |
|---|---|---|---|---|---|

21 September

| 2 | 6 | 4 | 8 | 5 | 6 |
|---|---|---|---|---|---|

**a** How many Units of electricity were used?

**b** How much did this cost if each Unit cost 15p?

**5 a** A 100 W light bulb is 4 per cent efficient. How much energy does it waste each second?

**b** A person changes their light bulbs for energy-efficient light bulbs, which are 24 per cent efficient. Explain how this may change their electricity bill.

**6** Calculate the frequency of a microwave. Its wavelength is $10^{-3}$ m (0.001 m).

The speed of light is 300 000 000 m/s.

**7** Use the equation: velocity = frequency × wavelength to fill in the table.

| Velocity (m/s) | Frequency (Hz) | Wavelength (m) |
|---|---|---|
| 330 | 30 | |
| 20 | 8 | |
| | 19 | 5 |

**8** Match the type of electromagnetic wave to its use.

| Type of wave | Use |
|---|---|
| Radio | In mobile phone networks |
| Microwave | In TV remote controls |
| Infrared | To broadcast programmes |

**9** Explain why microwaves are used to communicate with satellites.

**10** Match the type of electromagnetic wave to its use.

| Type of wave | Use |
|---|---|
| Ultraviolet | To kill cells in a tumour |
| X-ray | In sun beds |
| Gamma ray | To identify if a patient has cancer |

**11** Explain why gamma rays and X-rays are not suitable to use in the home.

**12** Write down three properties of all electromagnetic waves.

**13** What precautions should workers in X-ray departments take to reduce their risk from X-rays?

# AQA Examination-style questions

**1** Label each of the types of energy on this Sankey diagram for a television.

(1)

**2** Match the type of electromagnetic wave with its use.

| Type of wave | Use |
|---|---|
| Infrared | Security lights |
| Gamma | Mobile phone |
| Microwave | Sterilisation |
| Ultraviolet | Toaster |

(3)

**3** Choose words from the box to complete the sentences. Words can be used more than once or not at all.

> *energy   frequency   matter*
> *second   wavelength*

Waves transfer ............. from place to place without any ............. being transferred. If a wave has a small ............. it has a high ............. This means that more ............. is being transferred per ............. .

(3)

**4** Use the power label from a microwave oven to answer the questions. Show your workings as part of your answer.

> Microwave oven (Household)
> Model number: 196136
> Serial number: 7168301403
> Manufactured: November 2008
> Input: 1100 W    Output: 700 W

**a** Calculate how many units the microwave oven would use if left on for 15 minutes. (4)

**b** Calculate how much this would cost if the price per unit is 12 pence. (2)

**c** Calculate the efficiency of the microwave oven. (3)

**5** Read the article below from an internet site about skin cancer and then answer the questions.

> ### Cancer-beds outlawed for under 18s
>
> An expert committee that makes recommendations to the World Health Organization has suggested putting a ban on people under the age of 18 from using sunbeds.
>
> It made its decision following a review of research which concluded that the risk of melanoma – the most deadly form of skin cancer – was increased by 75% in people who started using sunbeds regularly before the age of 30.
>
> The Sunbed Association (TSA) supports a ban on under-16s, but argues there is no scientific evidence for a ban on young people aged 17 or 18.

**a** *In this question you will be assessed on using good English, organising information clearly and using specialist terms where appropriate.*

Explain the advantages and disadvantages to a person using a sun bed. (6)

**b** What could the government do to warn young people about the risks of using sunbeds? (1)

**1** **a** The letters represent the parts involved in a reflex action. Put them in the correct order in the flow chart.

................. $\longrightarrow$ ................. $\longrightarrow$ ................. $\longrightarrow$ .................

**A:** Relay neuron
**B:** Motor neuron
**C:** Receptor cell
**D:** Sensory neuron

(3)

**b** Explain the advantage of having a reflex action instead of a normal response. **[H]** (3)

**2** Choose words from the box to complete the sentences.

characteristics  generations  offspring  parents  years

Selective breeding involves selecting ............. with desired ............., crossing them then selecting the best ............. This process is repeated over several ............. (4)

**3** Sickle-cell anaemia is a common disease around the equator. A person with sickle-cell anaemia has red blood cells that are a different shape to normal. A husband and wife from Kenya wanted to start a family together but they were worried that their children might inherit the disease they were both carrying. Both of the parents had inherited one sickle-cell recessive gene and one normal dominant gene.

**a** Explain why neither parent shows the symptoms of sickle-cell anaemia. (1)

**b** Use a Punnett square to find the probability of these parents having a child with sickle-cell anaemia. Use **A** for the normal dominant gene and **a** for the recessive disease gene. **[H]** (3)

**c** Explain a problem there might be for someone suffering from sickle-cell anaemia. (2)

**4** A shop owner wanted to put the most efficient light bulb on display. She used a joulemeter to find the amount of energy that went into each bulb in a given time. She also measured the amount of light energy given off by each bulb in a given time.

**a** Complete her table of results

| Name | Energy needed by the bulb in joules | Light energy given from the bulb in joules | Efficiency (%) |
|---|---|---|---|
| Burnbright | 32 | 8 | |
| Lighteze | 20 | | 20 |
| Sureshine | 22 | 5.5 | |

(3)

**b** Explain which bulb would be the most expensive to leave on. (2)

**c** Explain why the bulbs are not 100 per cent efficient. (2)

**5** **a** The diagram shows a fuel-burning power station. Label the parts A to D (2)

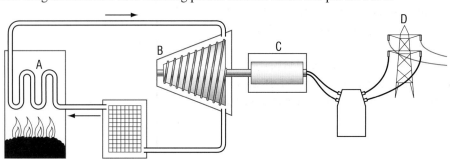

**b** Methane is a gas that can be burnt in a power station to generate electricity. Give a balanced symbol equation for the combustion of methane. **[H]** (4)

**6** Mortar is made from cement, water and sand.

**a** A builder used sand, water and cement to make up some mortar. He used $10\,cm^3$ of water and $20\,g$ of cement. He wanted to do an experiment to find how the amount of sand affects the strength of the mortar.

Give an outline of an experiment he could do to find the best composition for his mortar. *(3)*

The table shows a list of results that the builder found when he did his own more accurate.

| Amount of sand in grams | Breaking mass in kilograms |
|---|---|
| 5 | 10 |
| 10 | 20 |
| 15 | 35 |
| 20 | 30 |
| 25 | 26 |

**b** Write a conclusion based on the builder's results. *(2)*

**c** Suggest and explain what the builder could have done to make his experiment more accurate. *(1)*

**7** Read the newspaper article below about the opening of a coal mine and then answer the questions.

**Today the community of Lowfield was picketing the entrance of one of five sites identified by the government as possible locations for a nuclear power plant. Over 200 residents from the local community have decided to form an action group to oppose the site.**

The leader of the action group Josephine Lauder said 'There is no way this community is going to allow itself to be blighted so that the government can provide the country cheap electricity. We think the cost to our community is simply too high.' Miss Lauder's statements were in response to a report yesterday on a regional news programme. The report stated that not only will house prices dramatically decrease in the area but there is also scientific evidence that the power station can have a devastating effect on a community living in that area.

In response, the site manager looking at the location, Steven Ashman, adamantly denied the allegations made in the report and by Miss Lauder, and defied them to show him evidence that would back up their allegations. He also explained that the proposed plans would have an overwhelming positive effect on the community.

**a** State **two** reasons why the government wants to build more nuclear power plants. *(2)*

**b** *In this question you will be assessed on using good English, organising information clearly and using specialist terms where appropriate.*

The site manager, Steven Ashman, said that the power station would have a positive effect but people like Miss Josephine Lauder do not want to have a nuclear power plant built in their community. Give the reasons for and against having a nuclear power plant built close to a community. *(6)*

# Improving health and wellbeing

In Unit 3, Theme 1 you will work in the following contexts, covered in Chapters 10 and 11:

## The use (and misuse) of drugs

### What are medical drugs?

Medical drugs improve the quality of our lives by preventing, treating or curing diseases. These include antibiotics, such as penicillin, anti-inflammatory drugs like aspirin, and painkillers like paracetamol.

### What are recreational drugs?

Recreational drugs are taken for an individual's personal enjoyment. They have no medicinal benefits. In fact many cause more harm than good. Tobacco and alcohol are examples of legal recreational drugs. People often take them to relax. However, if they are taken over a long time they seriously damage your health.

## The use of vaccines

### What is a vaccination?

Medical scientists have developed vaccines against a number of diseases caused by bacteria and viruses. This results in a person developing immunity, and prevents them getting the disease. Vaccination is the simplest, most efficient and cost-effective way of preventing life-threatening infections in the community. Common vaccines include tetanus, polio and measles.

## How do our bodies prevent pathogens causing disease?

You come into contact with pathogenic microorganisms every day. Thankfully, they do not always make you ill. The skin acts as a barrier preventing the entry of the majority of these microorganisms. If the skin is damaged it must seal itself as quickly as possible. Your platelets play a major role in this process.

## The use of ionising radiation in medicine

### How is ionising radiation used in hospitals?

Medical professionals can diagnose and treat certain diseases using ionising radiation. For example, X-rays can detect broken bones, and gamma radiation can be used to treat some types of cancer. Gamma radiation is also used in the diagnosis of some medical disorders. There are many other medical uses of radiation that you will learn about in this theme.

### How is exposure to ionising radiation monitored?

Although ionising radiation has many medical uses, it is also potentially harmful to people who work with it every day. Radiation protection supervisors, medical staff, engineers and radiographers all work together to ensure that the received dose of radiation does not exceed safe limits. Workers who are exposed to radiation wear a film badge. This records the received dose of ionising radiation.

Matching patients to treatments by screening their genetic makeup is the goal of some drugs companies. The prescribing of drugs has always tried to tailor treatment to reduce side effects, but generally this relied very much on trial and error. Personalised medicine tries to predict treatment response or prevents disease before symptoms appear.

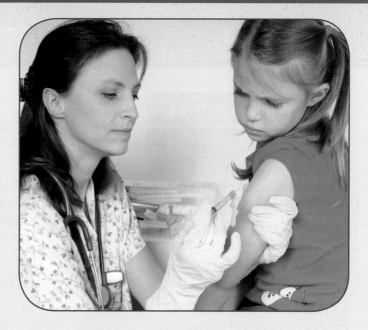

Drugs are chemical substances. They work by altering the chemical reactions that take place inside the body. If the body gets used to these changes, it may become dependent on a drug. This is known as addiction. Many highly addictive drugs are illegal in the UK; however, tobacco and alcohol are widely available to over-18s. Some medical drugs are also addictive – however, their usage is carefully monitored by doctors and medical professionals.

Before a drug can be prescribed by doctors it has to be carefully researched and monitored. Part of this process involves the new drug being tested on animals to ensure it is safe to use. Testing new drugs on animals is extremely controversial. Some people believe it is cruel, and a violation of the animal's rights. However, many people believe these tests are very important, as the drugs may have the potential to save human lives.

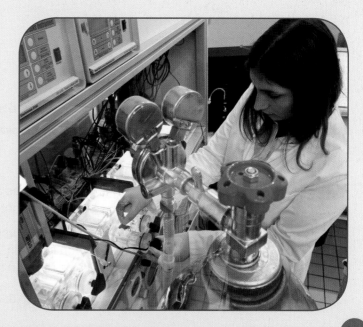

# 10.1

# Medicines

## Learning objectives

- What is a medical drug?
- What is the difference between a painkiller and an anti-inflammatory drug?
- What are some of the side effects of taking 'over the counter' medicines?

⊙⊙ links

*For more information on antibodies see 10.2 Antibiotics.*

?⁇? **Did you know … ?**

Aspirin has been developed from the bark of the willow tree. For thousands of years, ancient peoples used willow bark for pain relief.

From the Cinchona tree, we get quinine, which is used to treat malaria.

## Medical drugs

When you feel ill, doctors can prescribe drugs to make you feel better. These are known as **medical drugs**. These are legal, and help improve quality of life by curing, preventing or treating a disease.

Some drugs work by killing the microorganism that has made you ill. These microorganisms are known as **pathogens**. Antibiotics work in this way. Other drugs work by relieving the symptoms of an illness, like painkillers, sleeping tablets, high-blood-pressure tablets and antidepressants. These drugs do not provide a cure.

**a** What is a medical drug?

## Aspirin

Aspirin is a painkiller, often taken to relieve headaches. Drugs that reduce pain are also known as **analgesics**. Aspirin is also an **anti-inflammatory** – that means it reduces swelling – which in turn reduces pain. Headaches are caused when capillaries in the brain become inflamed. Aspirin does not kill pathogens.

**b** What type of drug is aspirin?

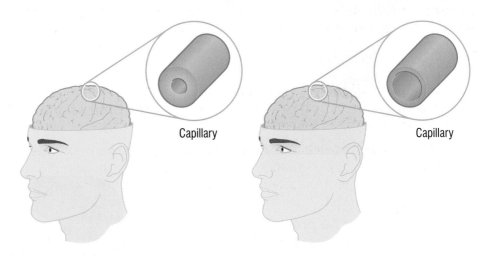

Before taking asprin          After taking asprin

**Figure 1** Action of aspirin

## Paracetamol

This drug is used as a painkiller. Like aspirin, it treats the symptoms of a disease but does not kill pathogens.

When you hurt yourself, your body releases a hormone-like substance that makes you feel pain. Paracetamol reduces the production of this chemical, decreasing the pain that you feel.

**c** What type of drug is paracetamol?

## Treatments for high blood pressure

Having high blood pressure puts strain on your heart and blood vessels. This increases the risk of heart attacks and other circulatory diseases.

If a person's blood pressure is slightly high, it can often be reduced by eating a healthier diet and exercising more. If it is dangerously high, medication will be prescribed. A range of drugs is available to reduce a person's blood pressure temporarily. These drugs act in a number of ways including:

● by reducing the amount of water in the blood
● by widening arteries, allowing blood to flow more freely.

These drugs do not provide a cure, they only relieve symptoms. Patients must take them constantly. If they stop, their high blood pressure will return.

**d** Why is high blood pressure dangerous?

## 'Over-the-counter' drugs

Many medicines are available 'over the counter'. This means people can buy them without consulting a doctor and getting a prescription. Many of these treatments are used to relieve aches, pains and itches. Others can prevent or cure ailments, such as athlete's foot. Some over-the-counter drugs help to manage recurring problems, such as migraines or period pain.

Generally, if you follow the instructions carefully, these drugs are very beneficial. However, long-term use or an overdose can have serious side effects. Interactions with other drugs or supplements can cause unexpected complications. People with other medical conditions, or those who are pregnant, should always seek the advice of a doctor before taking over-the-counter drugs.

**e** What is an 'over-the-counter' drug?

A number of problems can be caused by the over-use of medical drugs. The most severe conditions include kidney failure and liver damage. Anti-inflammatory drugs can also cause stomach ulcers, or problems with your intestines.

Many of these drugs cause serious complications if large quantities are taken in a short period of time – an overdose. This could result in a coma, or even death. Even if an overdose is not fatal, long-term complications can result. The long-term use of many drugs can also lead to addiction.

**Did you know ...?**

Analgesics are often used to reduce the symptoms of a headache. However, over-use of these drugs can trigger even more frequent headaches.

**Figure 2** Many drugs can be bought from supermarkets or chemists to treat common ailments. These are known as 'over the counter' medicines, and do not need to be prescribed.

### Key points

● Some medical drugs improve a person's health by curing, or preventing disease. Others relieve the symptoms of disease but do not cure it.

● Analgesic drugs reduce pain. Anti-inflammatory drugs reduce swelling, which may also relieve pain.

● Over-use of symptom-relieving drugs may lead to addiction, and can cause kidney and liver damage.

### Summary questions

1 Copy and complete the following sentences using the words below:
*inflammatory pain analgesic symptoms cure arteries swelling*

Some drugs treat the ............ of a disease, but do not provide a ............ Aspirin is an anti- ............ It is a drug that can relieve ............ Paracetamol is an ............, as it can reduce ............ Drugs that treat high blood pressure often widen ............

2 Explain why aspirin is described as an analgesic and an anti-inflammatory drug.

3 What are the potential health risks associated with over-use of paracetamol?

## 10.2

# Antibiotics ⓚ

### Learning objectives

- What is an antibiotic?
- Why should doctors be careful not to over-prescribe antibiotics?
- What is MRSA?
- How do antibiotic-resistant strains of bacteria develop? [H]

**Figure 1** This is the mould *Penicillium* growing on an orange

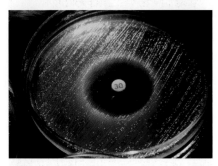

**Figure 2** Penicillin on an agar plate

### ∞ links

*For information about how bacteria evolve to become resistant to antibiotics look back at 3.5 Evolution.*

## What are antibiotics?

**Antibiotics** are drugs that kill *bacteria*, but do not damage the cells in your body. They have *no* effect on viruses or many fungi. Some antibiotics may have side effects that damage animal cells.

## Penicillin

Penicillin is an antibiotic. It is used to treat a range of conditions caused by bacteria, including ear, nose and throat infections.

Penicillin was discovered by accident! Back in 1928, pharmacologist Alexander Fleming was growing bacteria on agar plates. One day he forgot to seal one of the plates, leaving it open. When he returned he found a mould (called *Penicillium notatum*) growing.

He noticed that where the mould was growing the bacteria had died – something in the mould had killed them. The substance that had killed the bacteria was extracted and named penicillin.

**a** What do antibiotic drugs do?

**b** Name some common conditions penicillin can be used to treat.

The photo of an agar plate in Figure 2 shows where bacteria have been killed by penicillin. Notice the 'halo' around the penicillin disc. It is called the **zone of inhibition**.

**c** How can you tell from the appearance of an agar plate if bacteria have been killed by an antibiotic?

There are several different types of antibiotic. Each kills a different species or range of species of bacteria. To identify the type of bacteria that is making you ill, doctors may send samples, such as blood or urine, to be tested at public health laboratories in hospitals. Scientists grow the bacteria in the samples on agar plates so that they can be identified. The best antibiotic to kill the microorganism will then be prescribed to the patient.

### Practical

#### Antibiotics

Test the sensitivity of harmless species of bacteria to particular antibiotics using antibiotic rings. You could also investigate other materials that may have antibiotic properties – these could include tea tree oil, washing up liquid and toothpaste, for example.

## Antibiotic resistance

Bacterial infections have been treated with antibiotics ever since penicillin was discovered and developed. Since then antibiotics have saved millions of lives.

However, unfortunately, bacteria can spontaneously **mutate**, and these mutations can lead to some strains becoming **resistant** to antibiotics. This means that many types of antibiotic will no longer kill them. They are often referred to as 'super bugs'.

## How bacteria become resistant

When antibiotics are used to treat an infection, they will kill individual pathogens that do not have antibiotic resistance. However, resistant pathogens will survive. These will then reproduce, increasing the population of the resistant strain. Failure to complete a course of antibiotics 'because you feel better' also enables the more resistant strains to survive.

To try to slow down the rate of development of resistant strains, doctors no longer prescribe antibiotics for non-serious infections, such as mild throat infections.

## MRSA

MRSA is a bacterium that is resistant to many antibiotics. This means that MRSA infections are very difficult to treat and can be fatal. It is a particular problem in hospitals, because seriously ill patients are more likely to pick up this infection. However, the spread of MRSA can be limited through good hygiene practices.

**d** What does 'antibiotic resistance' mean?

## Over-prescribing antibiotics

Many people visit their doctor with minor illnesses, such as coughs and colds. In the past, doctors would nearly always prescribe antibiotics to treat these conditions. However, many of these conditions are caused by viruses – and so the antibiotics have no effect on the illness. The widespread use of antibiotics has increased the rate of development of antibiotic-resistant bacteria. These bacteria survive and reproduce, causing an increase in their numbers.

This creates extra costs for the National Health Service:

- more nurses and doctors are needed to treat new infections caused by the drug-resistant bacteria
- more staff are required to control outbreaks of these infections, and improve hygiene standards
- research into new drugs is expensive
- the new drugs themselves are expensive.

**e** Why shouldn't antibiotics be used routinely to treat minor infections?

### Summary questions

1 Copy and complete using the words below:

*antibiotic   bacteria   MRSA   penicillin   resistant*

............ is an ............ drug. It works by killing ............ . Some bacteria are now becoming ............ to antibiotics. One example is ............

2 Tonsillitis is a bacterial infection. It causes swelling in the tonsils, making the throat very sore. If you have it:
   **a** how can taking antibiotics make you better?
   **b** how can taking aspirin help relieve the symptoms?

3 How do antibiotic-resistant strains of bacteria develop? **[H]**

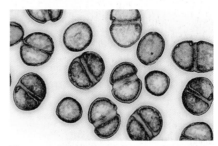

**Figure 3** MRSA bacteria

## 10.3

# Approving a new drug

### Learning objectives

- How is a new drug tested to see if it works?
- How do we make sure a new drug is safe?
- What are the issues surrounding testing drugs on animals?

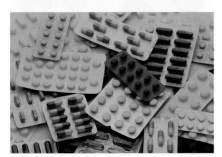

**Figure 1** New medicines improve our quality of life

### Did you know … ?

More than one in five of the world's top medicines were discovered and developed in the UK – more than the rest of Europe combined.

### AQA Examiner's tip

You need to know the steps involved in researching and testing a new drug. If you draw a flow chart of these steps, it may help you to remember them.

### Did you know … ?

Despite strict regulations, taking part in a clinical trial carries risks. Severe and unexpected side effects can occur.

### New drugs

Life expectancy in the UK is higher than it has ever been before. If you were born in 1900, you could have expected to live for 45 years. Now the average life expectancy is around 80 years.

Improving healthcare is one of the reasons people can now expect to live longer. Every year, many new drugs are developed by **pharmacologists** and put on sale in the UK.

### How are drugs developed?

New drugs come from a variety of different sources. Many come from plant extracts, and some are made chemically. Pharmacologists look for chemicals with interesting structures or properties. When they find a chemical they think might be useful, laboratory tests are performed to find out how it behaves. This is done before it is tested on living organisms.

### Testing the safety of new drugs

Medical drugs have to be tested to ensure they are safe and effective before they can be prescribed by doctors. The development of a new medical drug takes a long time and is very expensive. The flow diagram below shows the stages that are usually followed to test a drug:

| | |
|---|---|
| Drug is tested using computer models and human cells grown in the laboratory | Many drugs fail at this stage because they damage cells, or do not appear to work |
| Drug is tested on animals, such as nematode worms, fruit flies and mice. The animal is monitored closely for any side effects | In the UK, medicines are tested on animals, but it is illegal to test cosmetic and tobacco products on them |
| Drug is tested on a small group of healthy human volunteers to check it is safe | Testing drugs on humans is known as clinical trials |
| Drug is tested on volunteer patients who have the illness that is being targeted to ensure it works | Usually several hundred people |
| Drug is tested on patients to achieve data on drug effectiveness, safety, dosage and side effects | Usually several thousand people, to achieve reliable data |
| Drug is approved by the Medicines and Healthcare products Regulatory Agency (MHRA), and can be prescribed | |

**a** What does MHRA stand for?

**b** Why are new drugs tested on healthy volunteers before being tested on patients?

## Activity

### Can clinical trials go wrong?

TGN1412 was a new drug, intended for use in the treatment of leukaemia and arthritis. Tests on monkeys had been successful but its effects on humans were not known. The trial went horribly wrong and six of the volunteers nearly died.

Discuss whether new drugs should be tested on human volunteers.

## Activity

### Drug testing debate

Discuss the following:

- Do you think it is acceptable to test medical drugs on animals?
- Should testing never occur on animals?
- Do you think there are certain times when it is appropriate?

Even after a new drug is licensed, it is still closely monitored. This is in case there are unexpected side effects that have not been discovered. Sometimes, for instance, a drug might interact with other types of medication or illnesses. If there are reports of this happening, a drug might be withdrawn while more tests are carried out. Anyone in the UK can report a side effect to the MHRA.

### Why are some people against testing drugs on animals?

Some people feel very strongly about testing drugs on animals. However, all medical drugs were once tested on animals.

**Arguments for testing:**
- Drugs are developed that can save or improve people's lives.
- Animal testing generates information on how drugs affect a living body.
- Animal lives are not as valued as human lives.
- Animals have a shorter lifecycle than humans. So the long-term effects of a drug can be studied in a relatively short time.
- Many animals can be tested at one time.
- Testing animals is cheaper than carrying out research on humans.

**Arguments against testing:**
- Animals have the right to life; we should not experiment on them.
- Testing drugs on animals can cause pain or discomfort for the animal.
- Many animals die during the testing or have to be put down after the trial.
- The reaction of an animal to a drug may be different from that of a human.

## Key points

- Many new drugs come from plants and some are made synthetically.

- Drugs are tested in the laboratory, on animals and on human volunteers to make sure they are safe.

- Arguments for testing drugs on animals include: saves human lives; shorter life cycles of animals; cheaper than testing on humans.

- Arguments against testing drugs on animals include: the animals suffer; animals have the right to life; results may not be the same for humans.

## Summary questions

1 Copy and complete using the words below:

   *effects living animals volunteers tested thousands safe*

   New drugs are always ............ carefully in order to make sure they work and are ............ New drugs are tested on ............ to find out their effects on ............ organisms. Human ............ are given the new drug to find out if it causes side ............ Wide-scale trials then test the drug on ............ of patients before it is licensed.

2 Why is life expectancy in the UK higher now than it was 100 years ago?

3 Make a table listing the arguments for and against testing new drugs on animals. Explain whether or not you feel animal testing can be justified.

## 10.4 Recreational drugs

### Learning objectives

- How do drugs affect the body?
- What are recreational drugs?
- What is drug addiction?

### How do drugs affect us?

**Drugs** are chemical substances that affect the way our bodies work. Some have a helpful effect when we are ill. However, many can seriously harm our health and even cause death. **Recreational drugs** are taken for a person's enjoyment. They have no medical purpose.

Drugs work by altering the chemical reactions that take place inside the body. If the body gets used to these changes, it may become dependent on a drug. If this happens to someone, they become **addicted**. They start to crave the drug and feel that they can't survive without it.

If addicts attempt to stop taking a drug they suffer **withdrawal symptoms**. This happens because the body is no longer being provided with a chemical that it is used to having. It takes a long time for the body to get used to not having the chemical and for it to start working normally again. The symptoms can be very unpleasant and make it even harder to give up.

Every drug produces different amounts and types of withdrawal symptoms. Physical withdrawal symptoms include: headaches, sweating, palpitations, muscle tension, sickness and diarrhoea. Emotional withdrawal symptoms include: feeling panicked or irritable, being unable to sleep or concentrate and depression.

a  What sort of effects do drugs have on the body?

b  What happens if a person becomes addicted to a drug?

### Legal drugs

Some drugs that can harm our body are legal. These include alcohol, tobacco and prescription drugs, such as antidepressants and barbiturates.

- Many people think **alcohol** is perfectly safe, but it can seriously damage your body. It affects your nervous system and damages your liver.
- Most people realise **smoking** is dangerous, but around one in five of the population smoke. Cigarettes contain a drug called nicotine, which is very addictive. This makes it very difficult to stop smoking once you have started. Smoking seriously increases your risk of cancer and lung and heart diseases.
- **Antidepressants** are prescribed by doctors to help relieve depression. They provide patients with short-term benefits. However, if people use them for a long time, it can lead to addiction.
- **Barbiturates** are sedatives or tranquillisers and are present in some sleeping tablets prescribed by doctors. They are highly addictive and people can easily take an overdose. Long-term use can lead to depression.

c  What harmful effects does alcohol have on your body?

**Figure 1** These are examples of legal drugs

## Illegal drugs

Some drugs that can damage your body, even in very small amounts, have been made illegal by the government. Despite this, many young people will be offered these drugs and some may want to experiment with these deadly chemicals. Others may feel pressurised by their friends to join in.

Illegal drugs can be grouped together by their effect on the body. There are three main groups:

1  **Stimulants** – make you feel more alert, awake and generally happier. They work by speeding up the nervous system

2  **Depressants** – reduce feelings of stress and panic, which can make you feel more relaxed. They work by slowing down the activity of the nervous system.

3  **Hallucinogens** – alter what you see and hear. They work by interfering with normal brain function.

> **d**  Why are some drugs illegal?
>
> **e**  Name **three** groups of illegal drugs.

Illegal drugs can increase the risk of some diseases of the lungs and heart. Smoking cannabis, for instance, can increase the risk of lung disease. Cocaine tightens up blood vessels in the heart, which increases the risk of heart disease. One recent study of people admitted to Accident and Emergency units for chest pains found that 10 per cent of them had cocaine in their blood.

The table below shows some examples of illegal drugs and their effects on the body.

| Drug | Type | Harmful effect on the body |
|------|------|----------------------------|
| Barbiturates, e.g. dolls | Depressant | Addictive; hallucinations; heart attack |
| Heroin | Depressant | Addictive; risk of coma |
| Amphetamines, e.g. speed | Stimulant | Addictive; memory loss; increased blood pressure |
| Cocaine | Stimulant | Addictive; aggression; brain damage; damage to soft tissues of the nose |
| Cannabis | Hallucinogen | Addictive; mental health damage; bronchitis; lung cancer |

### Activity

**Is this 'cool'?**

This woman has died from taking drugs.

Discuss the following:

**a**  Why do people take drugs?

**b**  What short-term and long-term risks are involved in drug taking?

**c**  What effects can drugs have on your body and on your behaviour?

**d**  Photographs like the one above are used by the government and health professionals to try to shock young people and stop them from taking drugs. Do you feel that campaigns like this work?

**e**  How would you inform people of the dangers of taking drugs?

### Summary questions

1  Copy and complete using the words below:

   *addicted   withdrawal symptoms   unpleasant   drugs*

   Chemicals that affect the way your body works are called ............... If you take them too often you may become ............... When addicts stop taking drugs, they suffer ............... ............... These can be very ............... and make it harder to give up.

2  State **four** reasons why people may be tempted to take drugs.

3  When is a barbiturate a legal drug?

### Key points

● Drugs are chemicals that affect the body in a helpful or harmful way.

● Recreational drugs are taken for personal enjoyment. They have no medical benefits.

● People can become addicted to drugs and suffer withdrawal symptoms if they try to stop taking them.

## 10.5

## Tobacco

### What is in a cigarette?

Tobacco smoke contains over a thousand chemicals, many of which are harmful. Three examples are:

● **Tar** – this collects in the lungs when the smoke cools. It is a sticky black material, which irritates and narrows your airways. Some of the chemicals it contains cause cancer.

● **Nicotine** – this is the addictive drug in tobacco. It affects the nervous system. It also makes the heart beat faster, and narrows blood vessels.

● **Carbon monoxide** – this is a poisonous gas. It stops the blood from carrying as much oxygen as it should.

**a** Why is smoking addictive?

**b** Which part of cigarette smoke causes cancer?

### Activity

#### Passive smoking

Smokers are six times more likely to die prematurely than non-smokers. As well as putting their own lives at risk, they also endanger others. Other people breathing in the smoke (known as passive smoking) have an increased risk of developing circulatory and respiratory conditions.

Tobacco smoke seriously affects people with asthma. It also causes complications in pregnancy, and considerably increases the risk of sudden infant death syndrome (this used to be called cot death). 17 000 children under five are admitted to hospital every year in the UK for respiratory problems caused by parental smoking.

● Smoking has been banned in enclosed public spaces, such as cinemas and restaurants. Do you think this is a good idea?

● Do you think smoking should be banned anywhere else in public?

● Is there anything you think should be done to prevent sudden infant death syndrome?

### Smoking diseases

#### Diseases of the circulatory system

Three times as many smokers suffer from heart disease than non-smokers. Their arteries become narrowed as fatty deposits are left on artery walls This prevents blood flowing properly. Smokers are also at a higher risk of blood clots. This may result in total blockage of an artery, leading to a heart attack or stroke.

## Diseases of the respiratory system

Chemicals in tobacco smoke affect the alveoli – the air sacs in the lungs where gas exchange takes place. Smoke causes the walls of the alveoli to weaken and lose their flexibility so they do not inflate properly when the smoker inhales. They can also burst during coughing. This reduces the amount of oxygen that passes into the blood, leaving the person breathless. This disease is called emphysema.

Lung cancer is caused by chemicals in the tar found in lungs of smokers. Nine out of every ten lung cancer patients are smokers.

The cells lining your windpipe have tiny hair-like structures called cilia. These cells also produce mucus, which traps dirt and microorganisms. The cilia sweep the mucus out of the airways and it is swallowed into your stomach. This help to keep your airways clean. Chemicals in smoke paralyse the cilia so that, instead, the mucus flows into the lungs. This makes it hard to breathe, and can causes infection, such as bronchitis. Smokers have to cough this mucus up, which can damage the lungs further.

c   Name **three** diseases that smokers are more likely to suffer from than non-smokers.

d   How do cilia help to keep the lungs clean?

## How does carbon monoxide harm the body?

Oxygen is transported around your body by binding to haemoglobin, found inside red blood cells. But if there is carbon monoxide in your blood, it will bind to haemoglobin instead of oxygen. If this happens, the red blood cell cannot carry oxygen. So less oxygen is carried around the body. Burning cigarettes produce carbon monoxide, which smokers inhale, so they are getting less oxygen.

e   What is the role of haemoglobin in the body?

**Figure 1** A lung with a build up of tar as a result of smoking tobacco

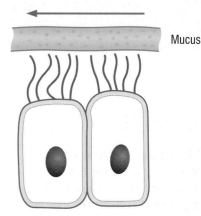

**Figure 2** A ciliated cell

### Activity

**Legal age**

The legal age to purchase cigarettes was raised from 16 to 18 in 2007. Why was this change introduced? Discuss with your group the arguments for and against making this change in the law.

### Key points

- The main components of tobacco smoke are tar, nicotine and carbon monoxide.

- Smoking increases your risk of suffering from circulatory and respiratory diseases.

- Carbon monoxide binds to haemoglobin in red blood cells. This lowers how much oxygen can be carried around the body.

### Summary questions

1   Match the contents of a cigarette to their harmful effect.

| Content of cigarette | Harmful effect |
|---|---|
| Tar | Addictive and makes the heart beat faster |
| Nicotine | Lowers the oxygen-carrying capacity of the blood |
| Carbon monoxide | Contains chemicals which cause cancer |

2   Smoking wastes money. If someone smokes 20 cigarettes at a cost of £5.50 a day, how much will they spend in a year?

3   Why do smokers often cough badly when they first wake in the morning?

# 10.6

# Alcohol

## Learning objectives

- How does alcohol affect the nervous system?
- How does alcohol damage the body?
- Is there a safe limit for drinking alcohol?

## Is alcohol a drug?

Many adults in this country drink alcohol, but this does not mean it is harmless. Alcohol contains the drug **ethanol**, which affects the nervous system. It is a depressant – it slows down your body's reactions.

Even in small quantities, drinking alcohol can change your behaviour. It causes people to lose their self-control. Most people feel relaxed and happy, but some can become aggressive or depressed.

>   **a** Why do many adults drink alcohol?
>
>   **b** What is the name of the chemical drug found in alcoholic drinks?

No alcohol

INCREASING INTAKE

**How a person would be affected**

Generally relaxed and happy
(over the legal limit for driving)

Dizzy and finding it difficult to walk as they lose control of their muscles – DRUNK

Slurred speech

Blurred vision

Unconsciousness

Death

Excessive alcohol

**Figure 1** What happens as people increase their intake of alcohol?

## How does alcohol affect your body?

When you drink alcohol, it is absorbed into your bloodstream from your intestines. It then travels to your brain, where it affects your nervous system. Alcohol can affect the body for several hours. It takes about an hour for the body to break down one unit of alcohol.

>   **c** How many small glasses of wine (125 ml) are equivalent to two pints of beer?

| One unit | One unit | One unit | One unit | One unit |
|---|---|---|---|---|
| 1/2 pint of beer | 1 small glass of wine | 1 single measure of whisky | 1 small sherry | 1 single measure of vodka |

**Figure 2** What is one unit of alcohol?

## Why is drinking large amounts of alcohol dangerous?

Heavy drinking over a long period of time can cause stomach ulcers, heart disease, and brain and liver damage.

The liver is responsible for breaking down alcohol in your body. It breaks down ethanol (which is poisonous) into harmless waste products. These are then excreted from the body.

The livers of heavy drinkers become scarred. Healthy cells are replaced with fat or fibrous tissue. The liver performs less efficiently. Then it takes much longer for alcohol and other toxins to be broken down. This condition is known as **cirrhosis** of the liver. This disease can be fatal.

>   **d** What happens to alcohol when it reaches the liver?

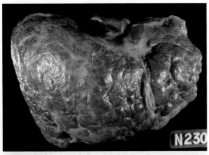

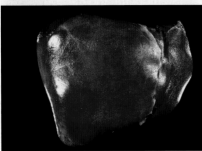

N230

**Figure 3** Compare these diseased (top) and healthy (bottom) livers

## Alcohol tolerance

When people drink alcohol regularly they need larger and larger amounts to have the same effect on their bodies. This is because their bodies have developed a tolerance to ethanol.

If they carry on drinking they may become addicted. They become dependent on alcohol and feel they cannot survive without a drink. These people are called alcoholics. Often, alcoholics do not realise they are addicted. Organisations such as Alcoholics Anonymous help people overcome their addiction.

**e** What is an alcoholic?

## Binge drinking

Government health advisers have set guidelines on alcohol intake. These are:

- adult males – 21 units per week (no more than four in one day)
- adult females – 14 units per week (no more than three in one day).

Some people exceed these guidelines and consume a large quantity of alcohol in a short period of time. This is known as binge drinking and, if excessive, can result in alcohol poisoning. This can be fatal. It also increases the risk of long-term health problems. Due to the effect of alcohol on the brain, it also often results in anti-social behaviour and crime.

**Figure 4** Binge drinking can result in long-term health problems and anti-social behaviour

---

### Activity

**Controlling binge drinking**

A number of factors have been blamed for the increase in binge drinking among young people in the UK. These include:

- alcoholic drinks to appeal specifically to younger people ('alcopops')
- wide availability of alcoholic drinks
- affordability of alcoholic drinks
- promotions, such as 'happy hours', that encourage young people to drink large quantities of alcohol in a short time
- extension of pub opening hours.

Some people have suggested that controls on alcohol should be put in place. These include the banning of 'happy hours', introducing a higher minimum price for alcoholic drinks, and ensuring advertising is not aimed at young people.

Discuss the advantages and disadvantages of introducing controls.

---

### Summary questions

1 Copy and complete the following sentences using the words below:

*nervous   relaxed   depressant   ethanol*

Alcohol is a ............ because it slows down your body's reactions. The drug in alcoholic drinks is called ............ It affects your ............ system. Small amounts usually leave the drinker feeling ............ and happy.

2 Explain why many heavy drinkers suffer from cirrhosis of the liver.

3 Explain why driving under the influence of alcohol is dangerous.

---

### Key points

- Alcohol is a depressant, which acts on the nervous system. It slows down the body's reactions.

- Long-term alcohol consumption can result in brain and liver damage.

- The government has set guidelines on the amount of alcohol a person should drink.

## Summary questions

**1 a** Select an example for each type of drug, below, by matching the letters and numbers.

| Type of drug | Name of drug |
|---|---|
| **A** Painkiller (analgesic) | **1** Penicillin |
| **B** Anti-inflammatory | **2** Aspirin |
| **C** Antibiotic | **3** Paracetamol |

**b** Which two types of drug, from the examples above, only treat the symptoms of a condition?

**2** Copy and complete the following sentences, using the words below:

*antibiotics   resistant   viruses   flu*
*tonsillitis   prescribed*

Some infections are caused by .............., such as .............. Bacterial infections, such as .............., can be treated using .............. These have to be .............. by a doctor. The over-use of antibiotics in the past has led to some bacteria, such as MRSA, becoming .............. to these drugs.

**3** Before a drug can be prescribed it has to undergo clinical trials. Reorganise the following steps of drug testing into the correct order.

**A** Drug is tested on patients to achieve data on drug effectiveness, safety, dosage and side effects

**B** Drug is tested on animals. The animal is monitored closely for any side effects.

**C** Drug is tested using computer models and human cells.

**D** Drug is approved by the Medicines and Healthcare products Regulatory Agency (MHRA), and can be prescribed.

**E** Drug is tested on a small group of healthy human volunteers to check it is safe.

**F** Drug is tested on volunteer patients who have the illness that is being targeted to ensure it works.

**4 a** What is the difference between a recreational drug and a medical drug?

**b** Copy and complete the following table, using these drugs:

barbiturates   cocaine   heroin   amphetamines

| Stimulant | Depressant |
|---|---|
|  |  |
|  |  |

**c** Why do people find it difficult to stop taking drugs?

**5**

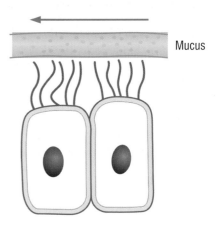

Mucus

**a** What is the name of this type of cell?
**b** What is its job in the respiratory system?
**c** How can this type of cell be damaged by smoking?

**6**

**a** What is the name of the drug in this picture?
**b** Which organ breaks down this drug?
**c** State two medical conditions that may be caused by the long-term use of this drug.

# AQA Examination-style questions

**1** Choose words from the box to answer the questions.

> aspirin  alcohol  antibiotic  cannabis
> heroin  nicotine  paracetamol

**a** Give the **two** legal medicines that are taken as a painkiller. *(2)*

**b** Which drug causes harm to the liver? *(1)*

**c** Which is the addictive drug in tobacco smoke? *(1)*

**d** Penicillin is an example of which type of medicine? *(1)*

**e** Which are the **two** illegal drugs? *(2)*

**2** Many microorganisms can cause diseases.

**a** Give examples of **two** diseases that are caused by a virus. *(2)*

**b** Describe how viruses make us feel ill. *(1)*

**c** Explain the role of platelets in stopping us from getting diseases. *(2)*

**d** Explain how our body becomes immune to a disease. *(3)*

**3** The table shows the effects of different types of performance-enhancing drugs used by sports men and women. Use the table to answer a–d.

| Performance-enhancing drug | Effect |
|---|---|
| Diuretic | Makes the person urinate |
| Lean mass builder | Grows muscle and reduces body fat |
| Painkiller | Allows a person to compete past their usual pain threshold |
| Sedative | Overcomes nervousness |
| Stimulant | Increases alertness |

**a** Explain the type of drug taken by a snooker player. *(2)*

**b** Which type of drug is caffeine? *(1)*

**c** Which type of drug might be taken by a wrestler to help him meet weight restrictions? *(1)*

**d** Which type of drug might be taken by a weight lifter? *(1)*

**4** Read these opinions about testing medicines on animals and then answer the questions.

> *A diabetic:* 'The insulin I have to take daily was tested on animals.'

> *A bypasser:* 'Drugs should be tested on prisoners who have committed serious crimes.'

> *A government official:* 'In Britain, the law states that if a drug is to be used for medicine, it has to be tested on at least two different types of live mammal.'

> *A Peta spokesperson:* 'The animals are routinely killed after they have been used in experiments.'

> *A news reporter:* 'Drugs testing on human tissue samples in a test tube and experiments carried out using computer models are both alternatives to animal testing.'

**a** Decide which people were giving fact or opinion.

| Fact | Opinion |
|---|---|
|  |  |
|  |  |
|  |  |

*(2)*

**b** Suggest why drugs are tested on animals before they go on sale. *(1)*

**c** Put the steps involved in research carried out in laboratories on drugs before they go on sale in the correct order. *(4)*

| A | B | C | D | E |
|---|---|---|---|---|
| Animals | Cells | Healthy humans | Tissues | Unhealthy humans |

*(4)*

**5** A study was completed in Bangladesh in 2004 to find if the number of years a person went to school for had any effect on them taking up smoking.

The graph shows the percentage of men and women smokers and how much schooling they received.

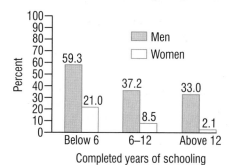

*In this question you will be assessed on using good English, organising information clearly and using specialist terms where appropriate.*

**a** Write a conclusion based on the results in the graph. *(6)*

**b** Explain why some people find it hard to give up smoking. *(2)*

# 11.1

# Harmful microorganisms

**Microorganisms** (also known as microbes) are very small living things – at least 100 times smaller than one of your cells. Most cause no harm to animals or plants. However, some can cause disease when they enter the body. These are called **pathogens** and are one of the things studied by microbiologists.

**a** What are pathogens?

**b** What name is given to scientists who study microorganisms?

There are a number of different groups of microorganisms. Most human pathogens belong to one of two groups, bacteria and viruses. These are studied in this spread.

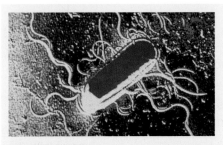

**Figure 1** Bacterium

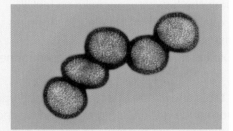

**Figure 2** Virus

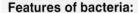

**Features of bacteria:**

- have a cell wall and cell membrane
- no nucleus
- genetic material floats around in the cytoplasm
- larger than a virus.

*Examples of bacterial diseases:*

- cholera
- typhoid
- tuberculosis.

**Features of a virus:**

- have a protein coat
- no nucleus
- a few genes that float around inside the virus
- smaller than bacteria.

*Examples of viral diseases:*

- measles
- rubella
- mumps
- polio.

**c** Give **two** differences between bacteria and viruses.

## How do pathogens make you feel ill?

Bacteria do this in two ways:

**1** They damage your cells. For example, lung tissue is destroyed by tuberculosis bacteria.
**2** They produce **toxins** – these are poisonous chemicals. *Salmonella* bacteria produce a toxin that causes one type of food poisoning.

Viruses make you feel unwell by getting inside and damaging your cells.

**d** How do viruses cause disease?

links
*For information about how doctors treat bacterial diseases through the use of antibiotics look back at 10.2 Antibiotics.*

## How do bacteria and viruses reproduce?

There is a time delay between harmful microorganisms entering your body, and you feeling unwell. This is called the **incubation period**. During this time, the microorganisms reproduce rapidly.

Bacteria can reproduce very quickly. In ideal conditions, they split into two every 20 minutes. So within hours, a few bacteria will have divided (replicated) into several thousand. They will now start to have a big effect on your body.

**e**  What happens during the incubation period of a disease?

Viruses cannot replicate by themselves. They can only reproduce by 'taking over' and using your cells (or those of another living organism) to make more virus particles.

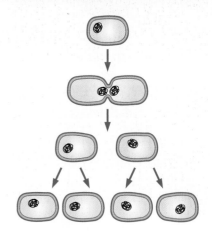

**Figure 3** Bacteria reproducing

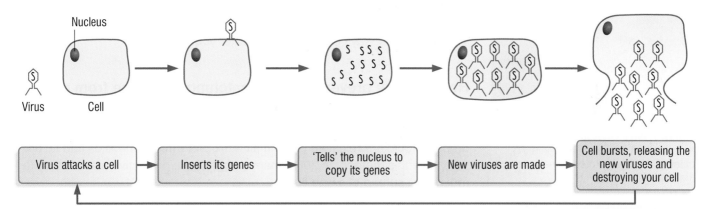

Nucleus

Virus    Cell

| Virus attacks a cell | → | Inserts its genes | → | 'Tells' the nucleus to copy its genes | → | New viruses are made | → | Cell bursts, releasing the new viruses and destroying your cell |

**Figure 4** Viruses must invade your cells in order to reproduce

## Summary questions

**1**  Copy and complete using these words:

*bacteria  mumps  pathogens  fungi  tuberculosis  viruses*

Microorganisms that cause disease are called ............. Three groups of microorganism are ..........., ........... and ............

An example of a disease caused by a virus is ............

An example of a disease caused by bacteria is ............

**2**  The table below shows the number of bacteria in a person.

| Time/minutes | Number of bacteria |
| --- | --- |
| 0 | 20 |
| 20 | 40 |
| 40 | 80 |
| 60 | |
| 80 | 320 |
| 100 | |

**a**  Copy and complete the table.
**b**  Plot a graph of the data.
**c**  Using the graph, explain why the person feels ill after a short period of time.

### Key points

- Bacteria, fungi and viruses are three groups of microorganism.

- Pathogens are microorganisms which cause a disease in a plant or an animal.

- Bacteria cause disease by damaging cells or producing toxins. Viruses cause disease by damaging cells.

# 11.2

# How are diseases spread?

## Learning objectives

● How do pathogens enter the body?

● How do diseases spread?

● How can you protect yourself from disease?

**Figure 1** What steps are taken to prevent the spread of disease in an operating theatre? Do these protect the surgeon or the patient?

**Figure 2** When someone sneezes, you can catch a disease through droplet infection

### ??? Did you know …?

Flies are often found on animal dung. They can then land on your food, spreading microbes all over it!

Restaurants and food outlets need to make sure all food is covered to prevent this happening.

In order to cause harm, pathogens have to enter our bodies. This can happen in four main ways:

● through cuts in the skin – from injury, or insect/animal bites
● through the digestive system – when you eat and drink
● through the respiratory system – when you breathe through your mouth and nose
● through the reproductive system – during sexual intercourse.

You are more likely to become unwell if large numbers of microorganisms enter your body. This is likely if you have contact with an infected person, or if you are exposed to unhygienic conditions.

**a** Name **four** ways pathogens can enter your body.

## How can you prevent yourself catching a disease from an infected person?

**By covering your mouth and nose**. When somebody sneezes or coughs, tiny drops of liquid are released into the air – **droplet infection**. Colds and flu are spread by this method. In 2009, many people wore protective masks to try to reduce the spread of swine flu.

**By not touching**. Some diseases are **contagious**. They are spread by touching infected people. Some can even be spread by touching objects an infected person has touched. Mumps and chicken pox can be spread in this manner.

**By using protection**. Body fluids are exchanged during sexual intercourse. Syphilis and gonorrhoea are examples of sexually transmitted diseases. Using condoms can help prevent diseases, including HIV, being transferred in this way.

**By not sharing needles**. People who inject medicines or illegal drugs should never share needles. Diseases can be passed on in blood on the needle. HIV and hepatitis can be spread in this way.

**b** How are colds and flu spread?

## How being hygienic can help prevent diseases spreading

**Cook food properly**. Some animals contain bacteria that could cause food poisoning. These include some species of *E. coli* and *Salmonella*. However, if the foods are cooked properly, the bacteria will be killed. You can also prevent the transmission of bacteria into cooked food by keeping raw and cooked food separate.

**Drink clean water**. Untreated water can contain microorganisms that cause diseases like cholera and typhoid. This may happen after flooding if sewage contaminates the fresh water supply. If water might be infected, you must boil it or use sterilisation tablets.

**Protect yourself from animal bites**. Mosquitoes spread malaria. By biting an infected person, and then someone else, the mosquito can pass a disease on. You can protect yourself from malaria by taking anti-malarial drugs, and using insect-repellent sprays.

**Wash your hands**. To reduce your risk of infection from microorganisms, wash your hands regularly. This is essential for employees in businesses involving the handling of foods.

**Cover cuts and grazes**. To prevent infection, you should thoroughly clean a cut. It should then be covered with a plaster, so microorganisms cannot enter your body.

c State **five** ways you can help prevent the spread of infections.

d How can insects transfer pathogens onto food?

An important part of many jobs is to ensure the health and safety of employees and clients. This includes ensuring that work premises are hygienic.

Restaurants are monitored by environmental health officers to ensure premises are clean, and that food is stored and cooked properly. This helps prevent the spread of microorganisms such as *Salmonella*, which could cause food poisoning.

Lifeguards are trained to check water purity in swimming pools and to use chemicals to ensure that water is clean. This prevents the spread of waterborne diseases.

It is very important that you maintain good personal hygiene at home to help prevent diseases spreading. This includes washing hands thoroughly before preparing and eating food. You should also ensure that you cover your mouth when you cough or sneeze. Ideally you should use a tissue, which should immediately be placed in a bin.

## Activity

### How can we prevent an epidemic?

'Swine flu' (Influenza H1N1) affected many countries across the world in 2009. Generally, it was a mild condition requiring painkillers, and a few days' rest. However, it was fatal for several thousand people around the world. To try to prevent its spread in the UK, the government:

1 Made the disease symptoms widely known – via the TV, radio and internet.
2 Promoted good hygiene procedures – for example, by encouraging the use of anti-viral handwashes.
3 Advised ill, elderly and pregnant people to avoid crowded places.
4 Shut schools where there were outbreaks.
5 Set up a swine flu helpline where symptoms could be checked over the phone.
6 Developed a vaccine against the virus that causes swine flu.
  Write a magazine article explaining how these steps reduced the spread of the disease.

**AQA Examiner's tip**

Make sure you read the question properly. There is usually a clue about the way the disease in question is transmitted. If you are asked for two ways to stop a disease being transmitted, make sure you do not give the same answer twice, e.g. 'cover mouth when sneezing' and 'cover mouth when coughing'.

## Summary questions

1 Copy and complete the table:

| Method of transmission | Diseases spread in this way |
|---|---|
| Shared needles | |
| Touch | |
| Droplets | |
| Sexual intercourse | |

2 Why must meat be thoroughly cooked before you eat it?

3 Why do diseases spread more easily in:
  a highly populated areas
  b countries with poor hygiene?

## Key points

- Pathogens enter the body through wounds (injury or bites), the digestive system, the respiratory system or during sexual intercourse.
- Diseases are spread by contact with an infected person, or exposure to unhygienic conditions.
- You can help to protect yourself from diseases by avoiding infected people, using condoms during sexual intercourse and by being hygienic.

# 11.3 Body defence mechanisms

## Learning objectives

- How does your body repair cuts and grazes?
- What is the difference between phagocytes and lymphocytes?
- What are antibodies?

Although you come into contact with harmful microorganisms every day, you are not always ill. The main barrier to infection is your skin. However, if your skin is cut or grazed, pathogens can enter your body.

To prevent pathogens entering your body, the skin needs to seal a cut as quickly as possible. This also stops you losing too much blood.

**a** What is your body's main defence against microorganism entry?

## How does the blood clot?

**Platelets** are small pieces of cell that are made in the bone marrow. They are carried around the body in plasma. These pieces of cell are essential for helping the blood to clot.

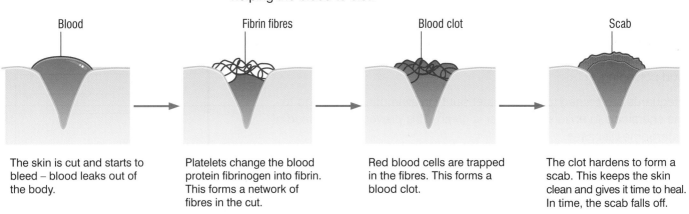

Blood | Fibrin fibres | Blood clot | Scab

The skin is cut and starts to bleed – blood leaks out of the body.

Platelets change the blood protein fibrinogen into fibrin. This forms a network of fibres in the cut.

Red blood cells are trapped in the fibres. This forms a blood clot.

The clot hardens to form a scab. This keeps the skin clean and gives it time to heal. In time, the scab falls off.

**Figure 1** How your blood clots

**b** How do platelets help the blood to clot?

## What happens if microorganisms enter our bodies?

Sometimes, harmful microorganisms do manage to enter our bodies. It is the job of white blood cells to prevent them causing disease. There are two types of white blood cells:

- **Phagocytes** – these cells engulf microorganisms.
- **Lymphocytes** – these cells make antibodies and anti-toxins.

## How do phagocytes fight disease?

Phagocytes **engulf** microorganisms. They then make enzymes which digest the microorganism.

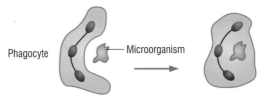

Phagocyte — Microorganism

**Figure 2** Phagocyte engulfing a pathogen

**c** How do phagocytes kill microorganisms?

### Did you know … ?

Leeches produce chemicals that stop the blood from clotting. These are used in hospitals to help heal sores, and in limb reattachment surgery. By stopping the blood clotting, oxygenated blood continuously enters the wound area (promoting healing) until the blood vessels regrow. Leeches also produce an anaesthetic, so you can't feel them sucking!

## How do lymphocytes fight disease?

Lymphocytes detect that something 'foreign' has entered your body. They then make an **antibody** – a chemical that attacks the microorganism.

The antibody reacts with the microorganism and de-activates it. Each antibody works for one type of microorganism. Every time a new type of microorganism enters the body, a different lymphocyte makes a new antibody to fight it.

After they have de-activated the disease, some antibodies remain in your blood. These antibodies prevent you getting the disease again. This provides you with **immunity**.

Lymphocytes also make **anti-toxins**. These chemicals destroy the poisonous toxins that some microorganisms make.

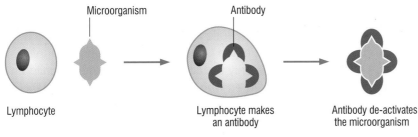

Lymphocyte    Microorganism    Antibody    Lymphocyte makes an antibody    Antibody de-activates the microorganism

**Figure 3** Lymphocyte making antibodies to de-activate a pathogen

**links**

*For more information about immunisation see 11.4 Immunisation.*

> **d** What are antibodies?
>
> **e** What are anti-toxins?

---

### Summary questions

**1** Match the parts of the blood with their function:

| Part of the blood | Function |
|---|---|
| Lymphocyte | Engulfs microorganisms |
| Platelet | Makes antibodies and anti-toxins |
| Phagocyte | Helps blood to clot |

**2** Why is it important not to knock a scab off a cut?

**3** What is the difference between an antibody and an anti-toxin?

**4** If you have had chicken pox as a child you should never suffer from the disease again. Explain why.

### Key points

- Platelets help the blood to clot. This hardens to form a scab, preventing microorganisms from entering the body.

- There are two types of white blood cells:

  - phagocytes, which engulf microorganisms

  - lymphocytes, which make antibodies and anti-toxins.

- Antibodies are chemicals that 'de-activate' microorganisms, stopping them from causing disease.

## 11.4

# Immunisation (k)

### Learning objectives

- How do you gain immunity to a disease?
- Why is immunity important?
- How do immunisations work?

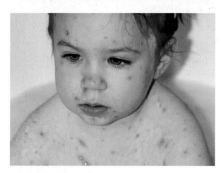

**Figure 1** The effects of chicken pox

### ?? Did you know …?

People often refer to immunisations as giving you artificial immunity. If you suffer from a disease and develop your own immunity to it, this is normally called natural immunity.

Both types of immunity develop as a result of your body producing antibodies. The major advantage of artificial immunity is you never have to suffer from the disease to start with!

### AQA Examiner's tip

A vaccine contains a small amount of the disease. The MMR jab contains small amounts of three diseases. If someone decided not to go for the MMR but had the three injections separately then each of the three injections would not contain a small amount of the MMR jab. Each of the three injections would contain a different disease.

When a pathogen enters your body, lymphocytes make an antibody against it. After the antibodies have destroyed the pathogens causing the disease, some remain in the body. If the same type of microorganism enters your body again, the antibodies will destroy it before it can cause disease. This is called **immunity**. It prevents you suffering from the same disease again.

### Activity

**Chicken pox vaccination**

Scientists have developed a vaccine against chicken pox, but do not routinely administer it. This is because in most cases chicken pox is a mild illness, and nearly 90 per cent of the population will develop immunity naturally. The immunisation is only usually offered to siblings of children with certain cancers, or of children who have had an organ transplant.

- Do you think the chicken pox immunisation should be routinely offered to every child? Carry out research using the Internet into the risks associated with chicken pox, shingles, and the vaccine, to make an informed decision.

**a** What is meant by the term 'immunity'?

## What is an immunisation?

Through studying immunity, medical scientists have developed **immunisations**. These are also referred to as vaccinations. Immunisations can protect you against some diseases caused by microorganisms. They are the simplest and most cost-effective means of preventing life-threatening infections in a population.

Are your immunisations up to date? Look at the table below:

| Child's age | Disease immunised against |
| --- | --- |
| 2, 3 and 4 months | Polio, diphtheria, tetanus, whooping cough, Hib meningitis and meningitis C |
| About 13 months | Measles, mumps and rubella (MMR) |
| 3–5 years | MMR, polio, diphtheria, tetanus and whooping cough |
| 10–14 years | Tuberculosis (TB) |
| 12–13 years | Cervical cancer (girls only) |
| 13–18 years | Polio, diphtheria and tetanus |

**b** Name **three** diseases you should be vaccinated against while you are at secondary school.

## How do immunisations work?

Immunisation involves a **vaccine** being inserted into your body. This normally takes the form of an injection but some vaccines can be taken by mouth.

Most vaccines contain dead microorganisms, or microorganisms that have been weakened so that they can no longer cause disease. The microorganisms do not make you ill but still trigger your white blood cells (lymphocytes) to make antibodies.

The antibodies destroy the microorganisms. Some antibodies remain in your body. These will be able to fight off the pathogen quickly if it enters your body again, preventing it from causing disease. You are now immune.

> **c** What does a vaccine normally contain?

## Diseases the whole population should be immunised against

### Polio

Polio is a disease that affects your nervous system. It can damage your nerves, and lead to permanent paralysis of parts of your body. In very severe cases it can cause death. The polio vaccine is not always injected as it can be given by mouth. It is often provided on a sugar lump as it doesn't taste very nice!

### Tuberculosis (TB)

You may recently have been injected with the BCG vaccine. This protects you against tuberculosis (TB). TB is a disease which affects your lungs and causes breathing problems. In very severe cases it can cause death.

### Measles, mumps and rubella

Just after their first birthday, children are given an immunisation called MMR. This single injection protects them against measles, mumps and rubella.

As well as being unpleasant, measles and mumps can cause permanent damage. In some cases, measles can cause deafness and mumps can make men infertile. In extreme cases, both diseases can be fatal.

Rubella is generally less unpleasant. However, if a woman gets the disease when she is pregnant, the unborn baby may be born deaf or blind. In some cases it can even die.

> **d** Why should children have the MMR immunisation?

### Summary questions

1 Copy and complete using these words:

*vaccines   immunity   diseases   microorganisms*

The spread of infectious ................. can be prevented by the use of ................. . These work by introducing dead or weakened ................. into the body. Your body builds up a defence to this infection – this is known as ................. .

2 Which diseases would a 15 year old have been immunised against?

3 Why do some parents choose not to have their children immunised against some diseases?

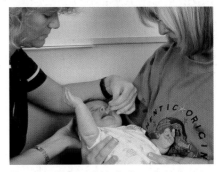

⌒⌒ links

*For information about how lymphocytes destroy microorganisms look back at 11.3 Body defence mechanisms.*

**Figure 2** This child is being given the polio vaccine

**Figure 3** The BCG vaccine is injected into the arm. This protects you from tuberculosis.

### Key points

- When a pathogen enters your body, lymphocytes make antibodies against it. Once these antibodies have got rid of the disease they remain in your body. This gives you immunity to the disease.

- Immunity prevents you suffering from the same disease again.

- Immunisations are doses of dead or weakened microorganisms. These are given by injection or by mouth. They trigger your body to produce antibodies. This gives you protection against a disease.

## 11.5 Vaccination issues

### Learning objectives

- How do we know that immunisations work?
- Do immunisations have side effects?
- Why do some people choose not to have immunisations?

The Government recommends that all children should have a number of immunisations during their childhood. These will protect them from many childhood diseases, and provide long-term immunity. If you travel abroad, other immunisations are sometimes recommended. This is especially true if you are visiting less-developed or tropical countries.

### How do we know that immunisations work?

During the period 1971–2000, the population of the UK increased. During this time, tuberculosis and measles injections were given to large numbers of the population. Despite the rise in population, the number of cases of these diseases decreased.

> **a** Explain why the number of cases of TB and measles fell during the period 1971–2000.

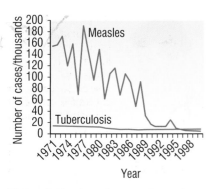

**Figure 1** Number of cases of measles and TB in the UK 1971–2000

### Why do some people choose not to be immunised?

Immunisations are an effective way of protecting yourself against many life-threatening conditions. Vaccines have to undergo many stages of testing, like all medical drugs. This ensures they are safe before they can be routinely administered to the population. However, not everyone chooses to be immunised. This may be because:

1 There are occasional scares about the safety of some vaccines.
2 Concerns over possible side effects.
3 Some people believe vaccines overload our immune system. They think this makes the immune system less able to react to other diseases such as meningitis, AIDS and cancer.

> **b** State **three** reasons why people may choose not to be immunised.

### What are the common side effects of immunisation?

Most people suffer few or no side effects from an immunisation. Any side effects are normally very short lived, and can easily be treated with a painkiller. The side effects that can be experienced include:

- fever
- sickness and/or diarrhoea
- swollen glands
- a small lump at the site of the injection, which may last for a few weeks
- irritability.

Severe reactions to immunisations are very rare.

**AQA** *Examiner's tip*

When answering questions referring to statistical data you should give figures from the graph to support your analysis.

∞ **links**

*For information on the many stages of testing that vaccines have to undergo look back at 10.3 Approving a new drug.*

## Do vaccines overload the immune system?

There is no evidence to suggest that having a number of vaccines, even several at a time, overloads the immune system.

Our immune systems are constantly being challenged by many different microorganisms. From the moment a baby is born, it is exposed to numerous bacteria and viruses on a daily basis. This is especially true whilst eating, and through putting their hands and objects into their mouth. Scientific studies have estimated that vaccines occupy less than 0.1 per cent of a child's immune system.

**c** Name **two** ways that babies naturally increase their exposure to bacteria and viruses.

## Activity

### Are vaccines effective?

Whooping cough is a disease which can cause long bouts of coughing and choking. This can make it hard to breathe. It can be fatal to babies younger than 1 year old. Children are now routinely vaccinated against whooping cough.

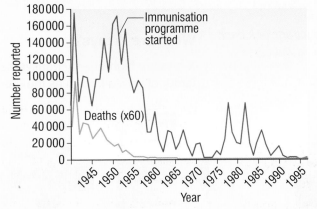

Graph showing the number of cases of whooping cough in the UK in the years 1940–1997. The graph also shows the number of deaths from this disease

### Questions

**a** In which year was the immunisation programme started?

**b** What effect has the vaccine had on the number of cases of whooping cough?

**c** How many deaths from whooping cough have occurred since 1980?

**d** Between 1972 and 1973, the vaccine was linked with Sudden Infant Death Syndrome. This link was later proved to be wrong. A lot of parents did not vaccinate their children in these years. What was the effect on the number of cases of whooping cough after this time?

**e** Suggest why the number of deaths from whooping cough was decreasing even before the vaccination programme was started.

## Summary questions

**1** Sort the following statements into reasons for and against immunisations:

- BCG leaves a scar
- Many lives are saved
- May get a temperature
- People don't suffer from infectious diseases
- Injections may hurt
- Side effects of disease may last for years, e.g. paralysis/deafness

**2** Why do scientists need to continually monitor the population for side effects caused by receiving immunisations?

**3** Some people cannot have immunisations, due to other medical conditions. Explain why immunising others helps to prevent these people from getting an infectious disease

## Key points

- Scientific data provides evidence that immunisations have decreased the number of cases of infectious diseases.

- Most people suffer no side effects from immunisation. If they do, they are generally very mild.

- Some people choose not to have immunisations because they are worried about side effects, the safety of vaccines and overloading their immune system.

# 11.6 Medical uses of X-rays

You may have had an **X-ray** photograph, perhaps to identify a broken bone. Have you ever asked yourself how an image of the inside of your body is made?

X-rays (a form of **ionising radiation**) were discovered over 100 years ago. Over time, scientists have discovered that ionising radiation can be very helpful to us, but it also kills living cells. Many of the early researchers working with X-rays developed forms of cancer.

Doctors and dentists use X-rays to see inside the body of a patient, without the need for an operation. This removes the chance of a patient developing an infection.

**a** Why do doctors take X-ray images?

## What are X-rays?

X-rays are high-energy transverse waves. They are just a small part of a large family of waves known as the **electromagnetic spectrum**.

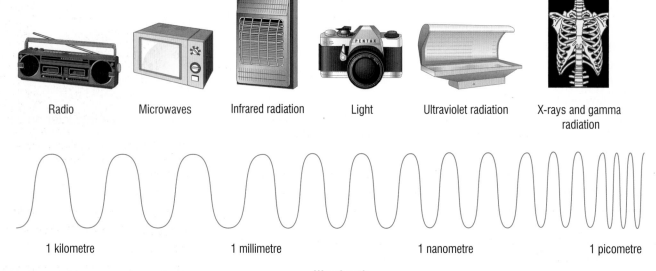

| Radio | Microwaves | Infrared radiation | Light | Ultraviolet radiation | X-rays and gamma radiation |

| 1 kilometre | 1 millimetre | 1 nanometre | 1 picometre |

**Wavelength**
(1 nanometre = 0.000 001 millimetres, 1 picometre = 0.001 nanometres)

**Figure 1** The position of X-rays within the electromagnetic spectrum

### links
*For information on the electromagnetic spectrum look back at 9.6 Electromagnetic waves.*

X-rays can penetrate through some materials. This means they can pass through them, without being absorbed. Dense materials absorb X-rays; this includes bones and teeth. The greater the density of the structure, the more absorption occurs. Modern X-ray imaging systems allow doctors to see fine detail in an X-ray image. This allows them to diagnose a wide range of medical conditions, including tumours and chest infections like pneumonia.

**b** What is an X-ray?

**c** State **three** medical conditions that can be diagnosed using X-ray images.

## How are X-ray images produced?

- Photographic film is placed behind the part of the patient being investigated.
- The X-ray generator is placed in front of the patient. The patient is exposed to X-rays.
- The X-rays penetrate through soft tissues like skin and muscle. They are absorbed by denser structures, such as bones and teeth.
- The X-rays that penetrate through the patient expose the film.
- The image is then developed. Regions of the film that were exposed to X-rays show up black. Areas of film that were not exposed to X-rays, because they were absorbed, are white.

Images produced in this manner are known as 'shadow pictures'. The image is formed because an X-ray shadow is produced behind dense material in the body.

d   Name some parts of the body which absorb X-rays well.

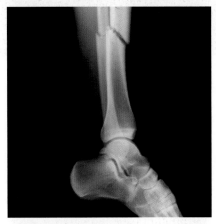

**Figure 2**  X-ray image of a broken leg

### Activity

#### Using X-rays to diagnose conditions

Doctors and radiologists use X-rays to diagnose a wide range of medical conditions. Carry out some research using the internet into which conditions can be diagnosed using this technique and those that cannot.

## How can we protect ourselves from the dangers of X-rays?

As X-rays are ionising radiation, they are known to cause cancer. However, the risk to your health of having a few X-rays is tiny.

Radiographers have to accept working with X-rays as part of their job. They are at risk of receiving high doses of this radiation. It is important that they protect themselves from this danger.

Lead (because it is very dense) absorbs X-rays, so a lead screen is placed between the radiographer and the patient. Radiographers also wear film badges. These measure the dose of radiation received.

e   How do radiographers monitor their exposure to X-radiation and protect themselves from the X-rays?

**Figure 3**  This radiographer wears a film badge to monitor her exposure to radiation

### ⬮⬮ links

*For more information on how film badges work see 11.7 What is radioactivity?*

### Key points

- Ionising radiation can treat cancer as it kills cells.
- X-rays are high energy, transverse, electromagnetic waves.
- 'Shadow pictures' are images of the inside of the body, which are produced using X-rays.
- Radiographers and medical professionals need to be protected from the harmful effects of X-rays.

### Summary questions

1 Copy and complete using these words:

*soft   waves   dense   inside*

X-rays are high energy ............ . They are used to make images of the .............. of the body. Only .............. parts of the body can be imaged, as X-rays will pass through .............. tissues.

2 Small children can have X-rays taken sitting on their parent's lap. Why does the parent need to wear a lead apron?

3 Explain why X-ray images are known as 'shadow pictures'. (Hint – think about how shadows are made using light.) Use a diagram to help with your explanation.

## 11.7 What is radioactivity?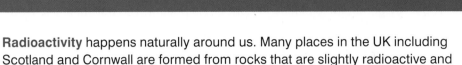

### Learning objectives

- What are alpha, beta and gamma radiation?
- How do they behave?
- How can we monitor a person's exposure to radiation?
- Why must their use be controlled?

**Figure 1** Rocks like granite give out a radioactive gas called radon

#### ⚬⚬ links

*For more information about radon gas see 15.9 Radon gas.*

#### ⁇⁇ Did you know ... ?

Radon gas is the second largest cause of lung cancer after smoking, although smoking causes at least 90 per cent of lung cancers.

#### ⚬⚬ links

*For information on atomic structure look back at 2.2 Inside atoms.*

**Radioactivity** happens naturally around us. Many places in the UK including Scotland and Cornwall are formed from rocks that are slightly radioactive and produce small amounts of radioactive radon gas. So what is radioactivity?

Nuclear radiation comes from inside the **nucleus** of some atoms. Some elements have unstable atoms that change over time. These atoms give out particles or electromagnetic waves from their nucleus and may turn into new elements. Three common types of radiation are:

- **alpha particles**
- **beta particles**
- **gamma radiation**.

Nuclear radiation **ionises** nearby atoms. The radiation knocks electrons from nearby atoms, leaving them positively charged. The electrons are added to other atoms, giving them a negative charge.

**a** What are the main types of nuclear radiation?

**b** What happens when an atom is ionised?

The table compares the properties of the different types of radiation:

| | Alpha particle | Beta particle | Gamma ray |
|---|---|---|---|
| | (+ +) | (−) | wwwwww |
| What is it? | Two protons and two neutrons | An electron | Very high energy, transverse electromagnetic wave |
| Mass compared with a proton | 4 | $\frac{1}{2000}$ | 0 |
| Charge compared with a proton | +2 | −1 | 0 |
| How ionising is it? | Very ionising | Weakly ionising | Very weakly ionising |
| Penetrating power | Absorbed by skin, thick paper or 10 cm of air | Passes through skin, thin aluminium or a few metres of air. Absorbed by lead. | Very penetrating. Partly absorbed by dense materials like lead and concrete. Thicker layers of these materials absorb gamma rays. |

c Which type(s) of radiation:

    **i** are stopped by lead,

    **ii** can pass through a hand,

    **iii** has the same charge as an electron,

    **iv** has a positive charge?

## The dangers of radioactivity

When molecules in cells are ionised by nuclear radiation, the cells can mutate, stop working properly or die. Radioactive radon gas can be harmful because it is breathed in. It can ionise lung cells, turning some of these cancerous. Many people in the UK have altered their homes so that radon gas cannot build up inside. They can install air pumps and ensure good ventilation.

Another way to protect people is to set a safe limit of radiation exposure. People working with radiation may be exposed to it over a long period of time. They often wear a film badge. These are checked regularly to measure their exposure to radiation. If the exposure is too high, they must stop working with radiation for a time. These badges do not protect workers but give information to help them limit their exposure.

d Write down **one** reason why we should reduce our exposure to radiation.

A film badge monitors a worker's exposure to different types of radiation. The badge contains a film covered with different coatings that turn black depending on how much radiation they are exposed to.

Each badge has windows covered by filters. The exact type of filter depends on the types of radiation a worker is exposed to. An open window, or plastic filters will monitor beta radiation exposure.

To monitor gamma radiation or X-rays, metal filters are used, usually tin or lead. The thickness of each filter determines exactly which type of radiation will get through to expose the film.

e Why do radiation workers need a film badge?

### Summary questions

1 Complete these sentences by choosing the right words:

  Alpha radiation is more dangerous inside/outside the body.

  Beta radiation comes from the nucleus/outer electron shell of an atom.

  Gamma radiation penetrates a long distance because it is strong/weak at ionising materials.

2 When the film from a badge is analysed, it is found that the film behind the metal filter is not affected. However, the film behind the plastic filter is black. Explain which type of radiation the badge has been exposed to.

3 Explain why gamma and beta radiation are more dangerous if the source is outside the body and why alpha radiation is more dangerous if the source is inside the body.

## ??? Did you know ...?

Our atmosphere absorbs gamma rays. This is why gamma radiation can only be studied using space-based telescopes.

## ⃝⃝ links

*The use of film badges to monitor radiation was introduced in 11.6 Medical uses of X-rays.*

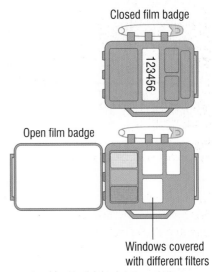

**Figure 2** The film badge monitors exposure to different types of radiation.

### Key points

- Nuclear radiation causes ionisation.
- Common forms of nuclear radiation are alpha, beta and gamma.
- Film badges can monitor the exposure of a person to radiation.

## 11.8 Uses of ionising radiation

### Learning objectives

- How is radioactivity used to treat patients with cancer?
- What are the ethical issues that should be considered when using radioactivity to treat patients with cancer?

### Gamma cameras

One way to investigate a patient who may have cancer uses **gamma cameras** and tracers. A tracer is a radioactive substance that is injected into the patient. The tracer travels through the bloodstream, and collects in the patient's bones or other tissues.

More of the tracer collects in regions where the cells are more active. This is where cancerous cells are repairing themselves. The tracer in the patient's tissues gives out gamma radiation. More intense gamma radiation is given out from places where the tracer collects. After a couple of hours, the patient passes through the centre of the gamma camera (see Figure 1).

The gamma camera detects where there is most radioactivity. This shows where the cancer is. Then doctors use this information to decide what treatment to give the patient.

There are some risks from using tracers and gamma cameras. The dose of radiation is equivalent to about 200 X-rays. This may be dangerous if the patient is a baby, a child or pregnant. The cells in very young children are more vulnerable to the effects of ionising radiation.

However, patients will only have a scan if doctors think they may have a serious health problem. The greatest risk is that cancer is not diagnosed and treated correctly.

a Write down one benefit and one problem for a patient having a gamma camera scan.

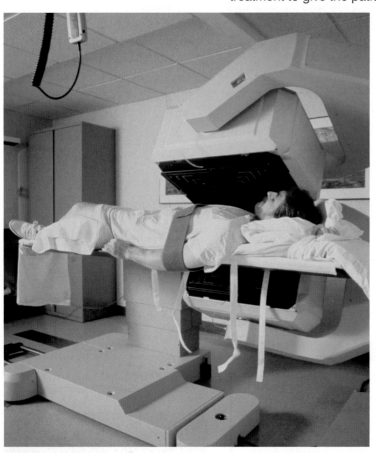

**Figure 1** A gamma camera creates a 3-D image of the patient

### Radiotherapy

Cancer cells can be killed using ionising radiation like X-rays or gamma rays. This is called **radiotherapy**. If cancerous cells are damaged or destroyed, they cannot divide so the tumour stops growing. Radiotherapy is also used after an operation to kill any remaining cancer cells. Cancerous cells are targeted to receive a large dose of radiation. Healthy cells are protected so that they receive a much lower dose.

In **external radiotherapy** ionising radiation is beamed at the tumour from outside the body. Doctors use several beams from different directions. This reduces the dose that healthy cells receive. Treatment takes a few minutes but is repeated over several days. Patients receiving external radiotherapy do not retain radioactivity as the radiation passes through them.

In **internal radiotherapy** a sample of radioactive material is placed next to the tumour inside the body. The sample emits gamma or beta radiation directly to the tumour. Cells further away receive a smaller dose. Less radiation is needed to treat the patient this way. Patients are radioactive during treatment because some radioactive material stays inside them.

**??? Did you know …?**

One source of radioisotopes used for medical treatment is from nuclear reactors in nuclear power stations.

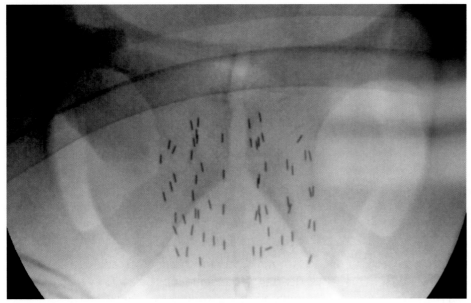

**Figure 2** Brachytherapy uses radioactive pellets placed next to the tumour inside the patient

## Ethical issues

Patients and doctors have to decide whether or not to use ionising radiation to diagnose or treat cancer. In many cases, a patient may be unwell during treatment but recover and live for many years afterwards. If the cancer can be diagnosed before it is advanced, patients are less likely to need surgery and they are more likely to be cured.

However, in some cases, the cancer may not be suitable for radiotherapy. Some patients may reject treatment if they are unlikely to recover or are worried about long-term damage from the treatment.

**b** What is radiotherapy?

### Summary questions

1 Copy and complete using the words below:

*internal tracers tumours*

............ are formed when cells divide uncontrollably.

............ are used with gamma cameras to diagnose cancer.

............ radiotherapy is used to treat cancer inside the body.

2 Explain **two** advantages of diagnosing cancer using a gamma camera.

3 Prepare a leaflet to explain to patients how a gamma camera helps to diagnose cancer and why it is used in hospitals.

4 Find out more about the different uses of tracers.

### Key points

- Radiotherapy is used to treat cancer.

- Gamma cameras and tracers can diagnose cancer.

- Radiotherapy is not suitable for treating all cancers.

- Patients and doctors need to balance the harm caused by receiving ionising radiation against the chances of successfully treating cancer.

# Summary questions

**1** Copy and complete the following table, summarising the characteristics of bacteria and viruses. The first line has been completed for you.

| Feature | Bacteria | Virus? |
|---|---|---|
| Cell contains cell wall | ✓ | ✗ |
| Replicate inside host cell | | |
| Cell has protein coat | | |
| Cell contains nucleus | | |
| Replicate by dividing in two | | |

**2** There are several steps you can take to avoid catching a disease. Match the disease, the method of spread of disease and the method of disease prevention.

| Disease | Method of spread of disease | Method of disease prevention |
|---|---|---|
| Syphilis | Touch infected person | Use barrier method of contraception, e.g. condom |
| Influenza (flu) | Sexually transmitted | Wear protective mask |
| Chicken pox | Droplet infection | Avoid infectious people |

**3** Produce a flow chart from the statements below, describing how the body repairs a cut or graze.
- The clot hardens to form a scab.
- Red blood cells are trapped in the fibres.
- The skin is cut, and blood starts to leak out of the body.
- Platelets change the blood protein fibrinogen into fibrin.
- This forms a network of fibres in the cut.
- This forms a blood clot.

**4** The following graph shows how the number of cases of tuberculosis (TB) varied in the United States between 1990 and 2006.

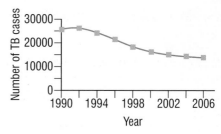

**a** Describe how the number of cases of TB in the US changed between 1990 and 2006.

**b** In which year do you think a TB vaccination programme may have been introduced? Use data from the graph to support your answer.

**c** Do you think the whole population had been vaccinated by 2006? Explain your answer.

**5** Rearrange the sentences below to explain how immunisations work.
- The antibodies destroy the microorganisms.
- These will fight the microorganism off quickly if it enters your body again, preventing it from causing disease.
- Vaccines contain dead or weakened microorganisms.
- Some antibodies remain in your body.
- The microorganisms do not make you ill but still trigger your white blood cells to make antibodies.

**6** Match each type of ionising radiation with its properties. Choose from these types of ionising radiation: alpha, beta or gamma.

**a** It is a form of electromagnetic radiation.

**b** It cannot penetrate through paper.

**c** It has a negative charge.

**d** It can penetrate through sheets of metal.

**7 a** Give **two** reasons why beams of gamma radiation outside the body are used to kill cancer cells inside the body.

**b** Write down **two** advantages of treating cancer this way.

# AQA Examination-style questions

**1** Your body has a defence system to get rid of any pathogens that enter.

Choose the correct word from each box to complete the sentences.

**a** There are two types of ......... that help to get rid of disease. *(1)*

> red blood cells    platelets    white blood cells

**b** These are the ......... that make antibodies that stick to the disease ... *(1)*

> lymphocytes    pathogens    plasma

**c** ... and the ......... that engulf anything with an antibody attached to it. *(1)*

> microbes    phagocytes    virus

**2** The graph shows the number of measles cases per year in Britain between 1996 and 2008.

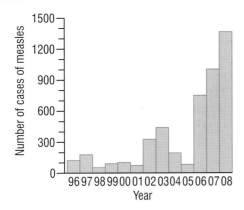

**a** In which year was the number of measles cases the lowest? *(1)*

**b** How many cases were there in 2003? *(1)*

**c** A triple vaccination called MMR is given to children under two.

What are the other two diseases the vaccination protects against? *(2)*

**d** Some parents want to give their child three separate vaccines rather than the triple vaccine MMR. Suggest **one** reason why this might increase the number of measles cases. *(1)*

**3** Radiographers can see inside people's bodies using X-rays.

**a** Put these sentences in the correct order to describe how X-ray images of broken bones are produced.

  **A** The X-rays that penetrate through the patient expose the film.

  **B** The part of the patient affected is exposed to X-rays.

  **C** Photographic film is placed behind the part of the patient being investigated.

  **D** X-rays penetrate through soft tissue like skin and muscle. They are absorbed by denser structures such as bones and teeth.

  **E** The image is then developed. Regions of film that were exposed to X-rays show up black. Areas of the film that were not exposed to X-rays because they were absorbed are white. *(4)*

**b** What type of wave is an X-ray? *(1)*

**c** Give the name and use of another electromagnetic wave that can be used in hospitals. *(1)*

**4** A film badge can be worn to monitor exposure to ionising radiation. The diagram shows a type of a film badge.

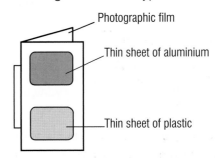

Photographic film

Thin sheet of aluminium

Thin sheet of plastic

**a** *In this question you will be assessed on using good English, organising information clearly and using specialist terms where appropriate.*

Describe how the film badge is used to detect the amount and type of radiation someone has been exposed to. *(6)*

**b** Suggest what disease someone who is exposed to too much radiation could get. *(1)*

# Making and improving products

In Unit 3, Theme 2 you will work in the following contexts, covered in Chapters 12 and 13:

## Uses of electroplating

Metals have many useful properties, making them tremendously important to us. However, some properties of certain metals make them less useful. For example, iron is prone to rusting and nickel can cause allergies. Electroplating is a way of using one metal to coat another. This can cover up a metal surface to stop it rusting, or to stop it coming into contact with skin.

Electroplating uses a process called electrolysis. This is when a compound is split up using an electric current. During electroplating, the object to be plated is put into a solution containing metal ions. The electric current makes the ions stick to the object and turn into atoms of the plating metal. In this way, the new metal builds up in layers.

Electroplating can be hazardous. Large electric currents can be dangerous and the chemicals used are often poisonous. People who electroplate metals need to minimise the risks they are exposed to. Safety equipment such as thick gloves and face masks are used to prevent workers from being harmed.

## Developing products

There are lots of new materials in everyday use because of the hard work of materials scientists. We now have:

- plastics that change colour as the temperature changes
- glasses, jewellery and clothes that change colour when they are exposed to light
- metals that always bend back to their original shape, or change shape when heated
- paints for cars that can heal their own scratches
- superconductors that conduct electricity without resistance and can make trains levitate.

Research into these materials is ongoing. Every year, more and more products that improve our quality of life are made.

## Selective breeding and genetic engineering

Fifty years ago, the human population was about 3 billion. Since then, it has more than doubled to about 6.8 billion and it is continuing to rise rapidly. As the number of people on the planet continues to increase, more food must be produced. Farmers and agricultural scientists try to solve this problem with selective breeding and genetic engineering.

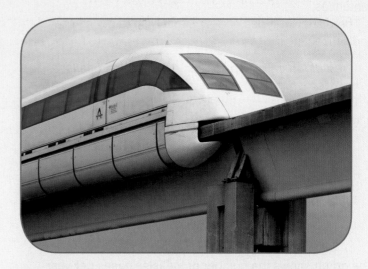

Selective breeding has been used for hundreds of years to make farm animals grow faster and larger. Only the animals with desired characteristics are allowed to breed. One of the problems of this is that it takes a long time to improve an organism through breeding. Another problem is inbreeding. When the breeding population is too small, genetic disorders can develop.

Genetic engineering involves changing the DNA of an organism to give it new or better characteristics. For example:

- crops that are resistant to disease or can grow in colder weather
- fruit and vegetables that contain extra vitamins
- bacteria that produce human insulin to treat diabetes patients.

Techniques like cloning and tissue culturing are other forms of biotechnology aimed at improving human life. Soon it may be possible to grow replacement organs for people, using just a few cells.

These earrings are made of nickel that has been electroplated with gold. The nickel is cheap and strong, but it is not very attractive and some people are allergic to it. Covering them with gold makes them look more attractive and prevents people developing allergies. It also helps the earrings last longer, as gold is less reactive than nickel and will not corrode.

Thermochromic pigments change colour when the temperature changes. For example, special inks have been developed that show when an egg has been cooked for the correct time. Customers can choose whether they want soft-, medium- or hard-boiled eggs, then buy eggs with a stamp that turns black after 3, 4 or 7 minutes respectively. Thermochromic road safety signs could be used in future to warn motorists about frosty weather conditions.

Superconductors are used in my MRI scanner. They are part of a very powerful electromagnet that is used to produce medical images. Without superconductors, it would be harder to detect brain tumours and other medical conditions. The superconductors allow electricity to go through them with no resistance. This means I can use very high currents to make strong magnetic fields. The superconductor needs to be very cold first, so liquid helium that is extracted from the atmosphere is used.

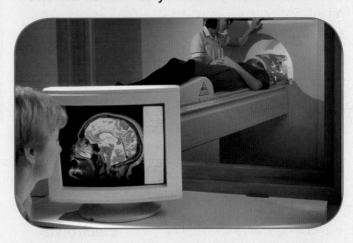

Researchers in Japan have added a gene to mice that makes them glow green. The gene was originally from a type of jellyfish that glows when exposed to blue light. The researchers did this in order to test different methods of genetic engineering. This kind of experiment improves scientists' understanding of genetic engineering. Greater understanding can lead to more ways to cure genetic diseases. Mice are not the only animals to have been genetically engineered in this way. Scientists have also created glowing pigs, monkeys, fish and even a rabbit. This type of research raises many ethical issues.

# 12.1 Electrolysis

## Learning objectives

- What is electrolysis and what can we electrolyse?

- How does electrolysis work?

There are lots of situations when chemists and materials scientists need to change a material chemically. It might be to make the material more useful, to make it last longer, or to give it new properties.

**Electrolysis** is a process that scientists can use to help cause a chemical change. It uses electricity. **Electrolysis is the breakdown of a compound by electricity**. During electrolysis, metals and gases may form at the electrodes. An example of the use of electrolysis is in **electroplating**. This is when one metal (or any material with a conducting surface) is plated (covered) with a thin layer of another metal by electrolysis.

> **a** What does electroplating mean?

## Performing electrolysis

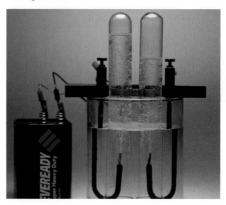

**Figure 1** Equipment for electrolysis

This photograph shows the equipment you need for electrolysis. The most important things you need are a source of electricity, **electrodes** and an **electrolyte**.

## Electrolyte

An electrolyte is the liquid that is being broken down. For a substance to be a good electrolyte, it needs to be able to split up into particles called **ions**. An ion is a charged particle and can be either positive or negative. Ions are formed when atoms gain or lose one or more electrons. Metal compounds split up into freely moving positive and negative ions when they dissolve in water or when they melt.

> **b** What is an ion?

## Electrodes

The electrodes are electrical conductors that dip into the electrolyte. They carry electrons to and from the electrolyte. Electrodes can be either positive or negative. This depends on how they are connected to the power source. The positive electrode is called the **anode**, and the negative electrode is called the **cathode**.

> **c** What is the difference between an anode and a cathode?

AQA Examiner's tip

You need to remember which is the positive electrode (anode) and which is the negative (cathode) as this might not be given in the exam question.

## Power source

The power source moves electrons to and from the electrodes that dip in the electrolyte. The current has to be **dc** (direct current, the one-way current you get from a cell), not **ac** (alternating current, the current supplied by the mains).

When the electrolyte conducts electricity, the positive and negative ions will move toward the electrodes. This happens because opposite charges attract. The positive ions move toward the negative electrode (cathode) and the negative ions move toward the positive electrode (anode).

Positive and negative ions are named after the electrode they are attracted to:

● positive ions are called **cations** (because they are attracted to the cathode)
● negative ions are called **anions** (because they are attracted to the anode).

An ion will be discharged (lose its charge) when it arrives at an electrode. This happens because it has gained or lost electrons. If the ion is a metal ion, then this will turn it back into a metal atom. This is how electroplating works.

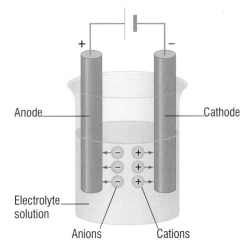

**Figure 2** Ions moving during electrolysis

⚭ **links**

*For more information on electrolysis see 12.2 Electroplating and 12.3 Reasons for electroplating objects.*
*For information about ions look back at 2.3 Different types of particles.*

**d** How can an ion become discharged?

### Summary questions

1 Match these words up with their definitions.

| | |
|---|---|
| Anode | Using electricity to break down a compound |
| Cathode | The negative electrode |
| Electrolyte | A charged particle |
| Ion | Using electrolysis to cover a metal with another metal |
| Electrolysis | The positive electrode |
| Electroplating | The liquid between the electrodes |

2 Why does an electrolyte need to be able to split up into ions?

3 Describe an anion and a cation without using the words positive or negative.

### Key points

● Electrolysis is the breakdown of a compound by electricity.

● Electrolysis involves passing an electric current through a liquid containing freely moving ions (which is called an electrolyte).

● Positive cations are attracted to the negative cathode; negative anions are attracted to the positive anode.

# 12.2 Electroplating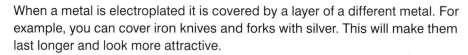

When a metal is electroplated it is covered by a layer of a different metal. For example, you can cover iron knives and forks with silver. This will make them last longer and look more attractive.

To do this, you could use silver nitrate solution as the electrolyte. When a product has been electroplated, it is usually called 'plated'. Here is how to do it:

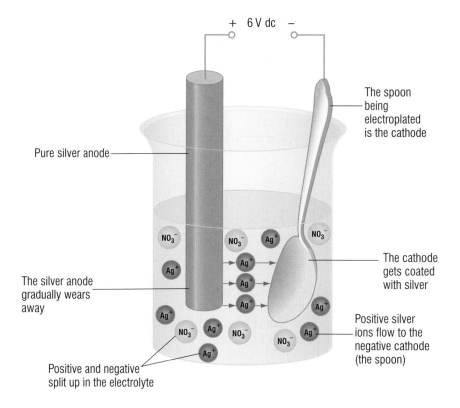

Figure 1 Electroplating silver onto an iron spoon

**a** Why would you electroplate silver onto knives and forks?

**b** What electrolyte could you use to electroplate silver onto an iron spoon?

The electrolyte is silver nitrate solution. Silver nitrate splits up into positive silver ions and negative nitrate ions in solution. Because of this, the silver ions are free to move around. Because they are positive, they are attracted to the cathode. This is because opposite charges attract each other.

As this happens, silver atoms from the anode are changed into silver ions. These dissolve in the electrolyte, replacing the ions that are coating the cathode. Gradually, pure silver starts to cover the iron spoon.

In practice, layers of other metals are electroplated on first. This helps the silver 'stick' to the object being electroplated. Objects plated like this are stamped EPNS, which stands for electroplated nickel silver.

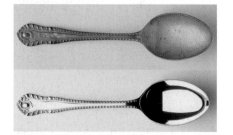

**Figure 2** Before and after silver plating

**c** Why are the silver ions attracted to the cathode?

## Practical

### Electroplating a coin

- How could you make a copper-plated 10p coin?
- What would you use as an electrolyte?
- What would the cathode and anode be made of?

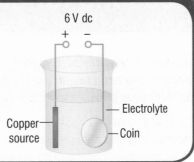

6 V dc

Copper source — Electrolyte — Coin

### ⭕⭕ links

*For more information on electroplating look back at 12.1 Electrolysis and see 12.3 Reasons for electroplating objects.*

**Higher**

## What is happening at the electrodes?

For the Higher Tier exam, you need to be able to write equations showing what happens at the anode and cathode.

### At the anode (+)

The cell removes electrons from atoms in the silver bar. This turns them into positive silver ions that dissolve in the electrolyte.

We can write:   $Ag \longrightarrow Ag^+ + e^-$

### At the cathode (–)

The cell gives electrons to positive silver ions arriving at the cathode. This turns them into silver atoms which plate the object.

We can write:   $Ag^+ + e^- \longrightarrow Ag$

To write electrode equations for different metals, you need to know the type of ion the metal makes. For example, copper sulfate can split up to make $Cu^{2+}$ ions. The copper ions have a charge of 2+. This means that each ion will have to gain two electrons at the cathode in order to turn into copper atoms.

This is how we would write the equations for an object being copper plated:

At the cathode: $Cu^{2+} + 2e^- \longrightarrow Cu$

At the anode:   $Cu \longrightarrow Cu^{2+} + 2e^-$

**d** Write electrode equations for electroplating of an object by zinc using a zinc anode and a solution of zinc sulfate.

## Summary questions

1 Copy and complete using the words below:

*anode electroplate cathode ions current plating move*

Electrolysis can be used to ............ metals onto other metals. The object being electroplated is used as the ............, and the ............ is made of the metal forming the plate. The electrolyte should contain ............ of the metal forming the plate. When the ............ is turned on, metal ions ............ from the anode to the cathode, ............ it in new metal.

2 Which electrode should be the object you are electroplating?

3 Look at the electrode equations for silver plating and copper plating. Which would require more electricity to deposit the same number of atoms on the cathode, and why?   **[H]**

## Key points

- Electrolysis can be used to plate metals with other metals in a process called electroplating.

- In electroplating, the cathode is coated with metal from the anode.

- We can write equations to describe what happens at the electrodes during electroplating. **[H]**

# 12.3 Reasons for electroplating objects

### Learning objectives

- Why do people electroplate metals and conducting surfaces?
- What precautions are taken to keep electroplating safe?

You would be surprised by the number of objects in your home that have been electroplated. Kettles, irons, parts of cars, jewellery, tin cans and even the coins in your pocket may have been electroplated.

This is normally done because another metal has more useful properties than the one the object is made of. Adding a very thin layer of the more useful metal gives the object many of its properties.

**a** Name **three** household items that are electroplated.

## Why metals are electroplated
### Preventing corrosion

**Figure 1** Parts of motorcycles are chrome plated

Many of the metal objects we use are made from iron or its alloy steel. This is because iron is cheap, easy to hammer into shape, strong and dense. The disadvantage of using iron is that it rusts easily. By electroplating the object, we can protect it from rusting but still use iron and benefit from iron's useful properties. A metal is used that is less reactive than iron, usually chromium. Chromium is used because it is hard, shiny and rust resistant.

Another use of electroplating to stop corrosion is in tin cans. Most cans of food are made from steel for the reasons given above. However, if the steel rusts the food will be ruined. Tin is quite unreactive, so it's an ideal choice to store food inside. A very thin layer of tin is used on the inside of the can. Cans plated like this are called tinplate steel. If a can gets dented, the tin layer can break. This exposes the steel, which can then start to corrode.

**b** Which metals are used to electroplate iron in order to stop it rusting?

**Figure 2** Tin cans are actually tin-plated steel

### ??? Did you know …?

Girls are about ten times as likely to be allergic to nickel as boys.

## Reducing the effects of allergies

Nickel is used in a lot of cheap jewellery, tools and fastenings on clothes. However, millions of people around the world are allergic to it. **Nickel allergy** can cause blisters, rashes and even sores on skin that comes into contact with the metal. By plating nickel jewellery with another unreactive metal, jewellers can keep using nickel without it harming the wearer. Usually, a metal like silver or gold is used as they are attractive and do not react with water.

> **c** What problems can nickel jewellery cause?

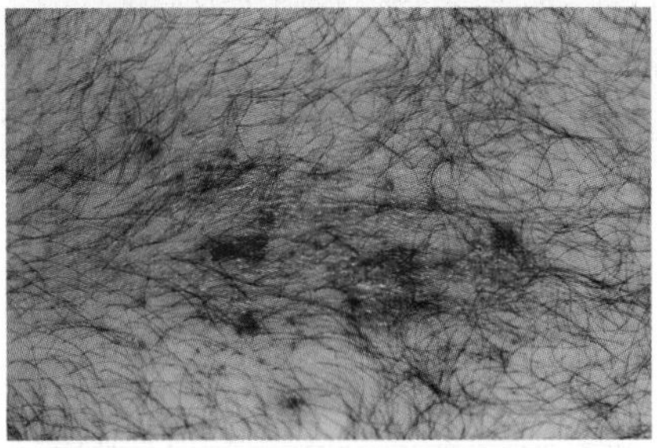

**Figure 3** Electroplating can protect people from allergies like this

## Decoration

Silver plating was described in the previous spread. Making objects like cutlery out of pure silver would be too expensive. Silver plating gives us the benefits of silver at a fraction of the cost.

## Electroplating safely

Many of the electrolytes used in electroplating are very harmful. For instance, an electrolyte often used in silver plating is silver cyanide. Cyanides are very toxic and can kill you within seconds if you inhale or even touch them. As well as posing a danger to those working as electroplaters, the toxic wastes may damage the environment.

There are lots of hazards in the workplace for an electroplater. Chemical burns, respiratory problems and poisoning are just some of the risks faced in electroplating. Working directly with electricity is also hazardous.

Electroplating companies use many precautionary measures to reduce the risk of staff being harmed. Safety boots, breathing masks, gloves, eye protection and chemical-resistant clothing are used every day. There are also strict rules about disposing of any chemical waste.

**Figure 4** Electroplated nickel silver

> **d** Name a poisonous chemical used in electroplating.

### Summary questions

1 Copy and complete using the words below:

*carefully hazardous chromium allergic nickel corroding safety tin*

............. is used to electroplate many iron items to make them rust resistant. ............. is used to plate iron in cans of food, to stop the iron ............. Cheap ............. jewellery is electroplated with metals like gold or silver to prevent ............. reactions. Electroplating can be ............. Workers must use ............. equipment at all times, and waste must be disposed of .............

2 Produce a summary table about electroplating using the information from this spread. Use the headings 'Item that is electroplated', 'Metal used', 'Reason for electroplating'.

3 List some of the hazards involved in electroplating, and describe the precautions taken to reduce the risks.

### ⦾ links

*For information on the properties of metals, look back at 7.4 Metals for construction.*

### Key points

- Objects are electroplated to make them resistant to corrosion, prevent allergic reactions and make them more attractive.

- The chemicals used by electroplaters are hazardous. Many safety procedures are in place to reduce the risk of harm.

# 12.4

# Developing smart materials

### Learning objectives

- What is a smart material?
- How do smart materials improve our lives?

Smart materials are materials that can change in response to their surroundings. More and more products are using smart materials because of their useful properties. Smart materials are used in clothes, medical devices, cars and many products you find in the home.

## Memory metals

Examples of smart materials are the shape memory alloys, or 'memory metals'. Memory metals are mixtures of metals that are 'set' into a particular shape when they are formed. They can be designed to have one shape when they are cold, and a different shape when they are hot. They can also be designed to spring back into their original shape if they get bent. They have lots of uses:

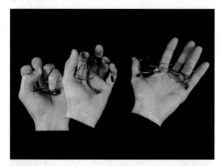

**Figure 1** Memory metal lets this frame spring back into shape

- **Spectacle frames** – they can be twisted and bent around without being permanently changed.
- **Braces** – the metal braces are 'set' to be in the correct positions for teeth and then attached to uneven teeth. Body heat makes the brace slowly contract back into its original shape. This corrects the teeth.
- **Medicine** – bundles of thin wire tubes, called stents, are put into narrowing blood vessels. Body heat makes them expand, holding the vessel open properly.
- **Robots** – memory metal can make 'muscle wires'. These expand and contract like human muscles when heated by a current.

> **a** How can memory metals be made to change back to their original shape?

## Self-healing paints

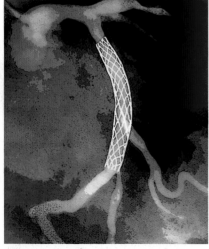

**Figure 2** Body heat is making this stent expand and hold a blood vessel open

Another group of smart materials are self-healing paints. They were first designed by car manufacturers, to reduce the appearance of scratches. Self-healing paints can be used on any painted surface that might get scratched. They work by releasing chemicals into the scratch after it is made. Some self-healing paints work like a very thick resin, slowly filling the scratch to make it disappear. They can make a scratch on a car door disappear within a week. Heat can speed the process up and warm water can help heal the scratch within a few minutes.

**Figure 3** Self-healing paint can make scratches disappear

Other self-healing paints are helped along by light. A new type of paint is based on a protein found in lobsters. Ultraviolet (UV) light makes chemicals in the scratch join together in chains, closing the scratch up.

**b** How can self-healing paints repair scratches?

## Chromic materials

Chromic materials have been around for a long time. They can change colour in different conditions. Materials that change colour at different temperatures are called thermochromic. Materials that change colour as a response to light are called photochromic.

**c** What is the difference between photochromic and thermochromic materials?

Thermochromic materials are usually painted in two layers. The top layer is thermochromic, and becomes transparent when heated. This reveals the layer underneath. Thermochromic materials can be used for:

● **Smart packaging** – if an item has been stored at the wrong temperature, the label will change colour so the user knows. This is useful for packaging medical supplies because certain medicines can be destroyed if they get too warm.
● **Safety materials** –spoons and bottles for babies can be made of thermochromic plastic. This changes colour if the food is too hot.

Photochromic materials have been used to make spectacle lenses for many years. In sunlight, photochromic lenses become tinted like sunglasses. When the wearer moves out of the light, the tint disappears. Other uses for photochromic materials are:

● **Rear-view mirrors in cars** – they darken at night so car headlights from behind don't dazzle the driver
● **Light detectors** – photochromic materials can absorb UV light, changing colour if the wearer has been in the sun for too long.

**d** Describe **three** ways chromic materials can stop people getting harmed.

**Figure 4** This spoon changes colour if the food is too hot

**Figure 5** A photochromic lens is only dark in bright sunlight

$AQA$ *Examiner's tip*

Make sure you know which applications are in the specification. These will not be given in the question, so you will have to remember them.

### Summary questions

1 Copy and complete using the words below:

*temperatures scratches change conditions heat colour memory healing*

Smart materials can ............ in response to different ............. Chromic materials change ............ as a response to light or ............ A ............ metal can change shape at different ............. Self-............ paints are able to fill in ............ on paintwork.

2 Make a summary table of the information on this page. Use the headings 'Type of smart material', 'Uses', 'How we benefit'.

3 Write a paragraph to describe the impact smart materials have had on medicine. Remember to consider social, environmental and economic implications.

### Key points

● Smart materials bring a wide variety of new properties to products. They can change depending on their environment.

● Smart materials include memory metals, self-healing paints, thermochromic materials and photochromic materials.

## 12.5 Superconductors

Some metals and **ceramics** have no **resistance** to electricity at very low temperatures. This is a very special electrical property called **superconductivity**. Resistance is a measure of how hard it is for an electric current to flow in any material. A current continues to flow in a superconductor even when the power supply is turned off because it has no resistance. Some superconducting wires have carried currents for years after the supply was turned off.

Superconductivity only occurs at temperatures so cold that the wires must be dipped in liquid helium or liquid nitrogen. Scientists are developing new materials that are superconductors at higher temperatures. So far, no material is a superconductor at room temperature.

**Figure 1** A superconductor can be made from ceramics. Ceramics are compounds heated to very high temperatures and then cooled. The superconducting material shown above is flexible, and can be moulded. When it has been heated it becomes hard, and stays in the required shape.

**a** What is the resistance of a superconductor?

**b** Why don't we see superconductivity in ordinary circuits?

Superconductors are used in electromagnets. An **electromagnet** is a coil of wire in a circuit. When the current flows in the wire, the electromagnet produces a magnetic field. If the current is turned off, the magnetic field disappears. If a superconductor is used as an electromagnet it stays magnetized even when the power supply is turned off.

Any wire with resistance transfers some electrical energy into energy heating the wire and its surroundings. In superconductors, the current can travel any distance with no energy wasted in the wire.

**c** How does an electromagnet made from superconductors behave differently from other electromagnets?

**d** How is the energy wasted in a superconductor different from that in normal conductors?

**Figure 2** Two sets of magnets in the track levitate the train and force it to move

# Where are superconductors used?

The uses of superconductors are limited by the need to cool wires to very low temperatures.

**Maglev trains** in Japan use magnets to make the train levitate and move. The train has superconducting magnets in its undercarriage, and there are electromagnets in the track's sidewalls. These sets of magnets repel each other so the train floats above the track. This reduces friction so the train can travel at up to 500 km per hour.

Another set of electromagnets in the track and train make it move. When the train passes coils of wire in the track's sidewalls, magnetic fields are created. These interact with the train's magnets to pull the train along at high speeds.

The superconducting magnets work even when their electricity supply is turned off. This way energy is saved, although it can be expensive to cool the magnets to the low temperatures needed.

**e** What are the advantages of Maglev trains?

## Medical scanning

**Magnetic resonance imaging (MRI scan)** uses pulses of radio waves to look inside your body. They build up a very detailed 2-D or 3-D image of tissues to investigate injuries and illnesses. Very strong magnets are needed for clear images – up to 40 000 times stronger than the Earth's magnetic field. The most common way to make magnets small enough and strong enough is to use superconductors.

The superconducting wire coil surrounds the tube that patients pass through. The coil is bathed in liquid helium at incredibly low temperatures. The helium is insulated in a container similar to a vacuum flask. This means the patient does not feel cold and the helium does not absorb energy from the surroundings. By using superconductors, the amount of electricity needed to run the MRI scanner is much lower than using other methods.

**f** Why are superconductors used in MRI scanners?

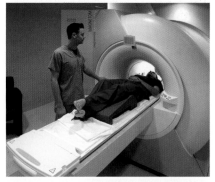

### ??? Did you know …?

Superconductivity was discovered by a Dutch physicist, Heike Onnes, in 1911. He investigated how different metals behaved at very low temperatures and used a mercury wire in liquid helium.

**Figure 3** Superconductors surround patients as they pass through the MRI scanner

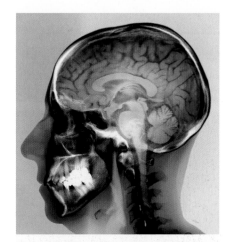

**Figure 4** MRI scans are very detailed

### Key points

- Superconductors have no electrical resistance at very low temperatures.
- They can make very strong electromagnets.
- They are used in MRI scanners and Maglev trains.

## Summary questions

**1** Copy and complete using these words:

*electricity   energy   resistance*

Superconductors have no ............ at very low temperatures. This means that there are no ............ losses when the current flows. The amount of ............ needed to make a very strong electromagnet using superconductors is quite low.

**2** Write down **two** advantages and **two** disadvantages of using superconductors in circuits.

**3** Find out one other use of superconductors and write a paragraph explaining how the superconductor is used.

**4** Some power stations use superconductors in their generators. These generators are about half the size of normal generators and over 99 per cent efficient. Why do you think some governments are funding research into superconducting generators?

<table>
</table>

# 12.6

# Advantages and disadvantages of modern products

## Smart materials

Many young people wear dental braces to straighten their teeth. Nowadays, some people use 'invisible' braces which are more comfortable and less noticeable than metal braces. However, these braces are expensive and need to be replaced every two weeks. They can only be used for teeth that are not too far out of place. The invisible braces are made from a modern plastic material that is transparent and strong.

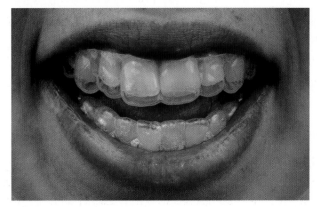

**Figure 1** Modern braces can be removable and invisible

**a** Write down **two** advantages and **two** disadvantages of 'invisible' braces.

It can cost hundreds or thousands of pounds to repair damage caused by scratching a car's paintwork. **Self-healing paint** is now used on some cars. This saves money on upkeep costs. The Sun's warmth or warm water allow the paint to reform after it has been scratched. The paint can be used on many hard surfaces and has been licensed for use on cars and mobile phones. It is more expensive than ordinary paint so it isn't used on low-cost items like toys. Its appearance is not suitable for some uses, such as painting rooms, and it is not available to use at home yet.

**b** Write down **two** advantages and **two** disadvantages of self-healing paint.

**Figure 2** Scratches on this car heal themselves

## Superconductors

The National Grid transmits electricity across the UK. If the wires were made from superconductors instead of aluminium, the savings would be enormous. Electricity could be transmitted at low, safer voltages and less energy would be wasted heating the conductor and its surroundings. However, no superconductor works at room temperature.

They must be cooled to very low temperatures. This makes superconductors very expensive to use and completely impractical for the National Grid. Superconductors are used to make very strong and compact electromagnets for MRI scanners and Maglev trains. They work even when their electricity supply is turned off, saving some energy.

**c** Write down **two** advantages and **two** disadvantages of superconductors.

**Figure 3** Superconductors only work at very cold temperatures so aluminium is used as the conductor in power lines in the National Grid instead

## Chromic materials

Thermochromic and photochromic paints change colour according to the temperature and light intensity. They are available for use at home on different surfaces, but are expensive compared with traditional paints. There is a limited range of colours and they may need to be covered with a clear layer of varnish to protect the coatings. Many of these paints are toxic and difficult to use. Also, they can fade if exposed to high temperatures, some solvents or UV radiation. They are not water based so their use and disposal has a greater impact on the environment than some traditional paints. However, using these paints can give more interesting results than using traditional paints.

> **d** Write down **two** advantages and **two** disadvantages of thermochromic and photochromic paints.

**AQA** *Examiner's tip*

Questions on this topic will often be comprehension or data-analysis type questions, so make sure you read the question properly before you answer anything.

### Summary questions

1 Copy and complete using these words:

*environment   expensive   money   variety*

Many modern materials are more ............. than traditional materials. Many modern materials save ............. by reducing the costs of repair or upkeep. Modern materials give people more ............. in the way they use different products. Using modern materials may affect the ............. more than traditional products.

2 Some hospitals are using special casts on patients after operations on their ankles and feet. The materials cost more but the patients don't need to stay in hospital as long. Write down **one** advantage and **one** disadvantage of this for the hospital and **one** advantage and **one** disadvantage for the patient.

3 Find out one other use of a modern material and write a paragraph explaining its advantages and disadvantages.

### Key points

- New and exciting materials are being developed by research scientists for new technologies or to replace traditional materials.
- Many modern materials are often more expensive than traditional materials.
- Modern materials are often more specialised than traditional materials.
- Producing and disposing of modern materials may affect the environment more than traditional materials.

# Summary questions

**1** To perform electrolysis, two electrodes must be connected to a power source. The electrodes are placed in a liquid called an electrolyte.
  **a** What are the two electrodes called? Which is positive and which is negative?
  **b** Explain what happens in the electrolyte when the current is switched on.

**2** During electroplating, metal ions are turned into solid metal.
  **a** Which electrode is the solid metal formed on?
  **b** Which electrode provides a source of metal ions?
  **c** What happens to $Cu^{2+}$ ions to turn them into copper metal?

**3** Many metals are electroplated to make them more useful.
  **a** What metal are cans of food electroplated with and why?
  **b** Why is nickel jewellery often plated with gold or silver?
  **c** Electroplaters must wear a lot of protective clothing at work. Name two hazards they protect themselves from at work.

**4** Complete these sentences using your own words:

  Superconductors have ............ resistance at very low temperatures. They are used to make electromagnets which stay magnetised when the ............ is turned off. These electromagnets are used in MRI scanners because they are very ............ for their size.

  One other use of superconductors is ..............

**5** A bracelet is made of a type of plastic that changes colour under UV light.
  **a** What is the name for this type of material?
  **b** How would the bracelet be useful?

**6** Materials scientists are constantly trying to improve products.

  Teflon is a smart material. It is the trade name for a polymer called polytetrafluoroethene or PTFE. It has many applications as it is very slippery and unreactive.

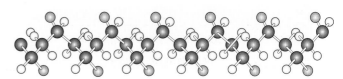

**A section of a Teflon® molecule**

> *Dirt will not stick to it   It has a high melting point   It is unreactive   It insulates It prevents allergic reactions It strengthens a structure*

  **a** Choose the reason why Teflon is used in clothing.
  **b** Choose the reason why Teflon is used as a coating for saucepans.
  **c** Choose the reason why Teflon is used to make containers for chemicals.

# AQA Examination-style questions

**1** Some metals need to be coated to make them more durable.

**a** Choose the correct word from each box to complete the sentences.

**i** Charged particles called .............. are made by taking away or adding .............. to atoms or molecules.

> atoms   electrons   ions
> charges   protons

(2)

**ii** .............. involves the movement of charged particles in an ...............

> charging   combustion   electrolysis
> air   an electrolyte   water

(2)

**iii** One use for electroplating is to ...............

> insulate
> prevent allergic reactions
> strengthen a structure

(1)

**b** During electroplating, the article to be coated is made the cathode. Explain what happens at the cathode when the article is being coated. **[H]** (3)

**c** A company wanted to speed up the process of coating metals. The graph shows the results of one of their investigations.

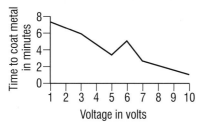

**i** Which voltage gave an anomalous result? (1)

**ii** Suggest what time the anomalous result should have been. (1)

**iii** Suggest what could have happened to cause the anomalous result. (1)

**iv** Write a conclusion based on the results in the graph. (1)

**2** Opticians have been experimenting with different materials to make spectacles better.

**a** Smart materials have been used to make the frames. Describe the use of smart materials to make spectacle frames. (2)

**b** Chromic materials have been used to make the spectacle lenses. What is the advantage of making the lenses out of photochromic material? (2)

**3** A medical company wanted to find the best superconductor to use in an MRI scanner.

**a** What is a superconductor? (2)

A multimeter is a device that can be used to measure the resistance of a device in an electrical circuit.

The diagram shows some of the apparatus used to find the best superconductor.

Superconductor

Multimeter

**b** What other variable would need to be measured? (1)

**c** *In this question you will be assessed on using good English, organising information clearly and using specialist terms where appropriate.*

Outline the method that the medical company could use to find the best superconductor in an MRI scan. (6)

**4** Read the article about biosensors before answering the questions.

### Smart wound dressings

Biosensors can measure the healing process of wounds. Scientists have made a wound dressing that can detect if anything goes wrong during the healing process.

The sensors can detect the pH value indicating the acidity of the wound, and a protein present when the wound becomes inflamed. These things can be detected quickly without needing to take the wound dressing off and allowing infections in.

Evaluate the cost implications for the NHS of using these new biosensor wound dressings compared with traditional dressings (3)

# 13.1 Selective breeding

⚭ links

*For information about natural selection look back at 3.5 Evolution.*

The farmer selects the ewe with the longest fleece; he breeds this with his best long-fleeced ram

The farmer chooses the best ewe and breeds again with his best ram

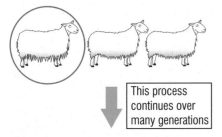

This process continues over many generations

Eventually all the sheep have the desired characteristic of a good-quality, long fleece

**Figure 1** Selective breeding of sheep for long wool

## What is selective breeding?

To ensure that they maintain or improve their stock, farmers choose to breed from their best animals or plants. This is called **selective breeding**. It is something that has been carried out for thousands of years, ever since people first began to farm.

Farmers will select the animals they breed from, or the plants they grow, by characteristics that are of benefit. For example, they may choose:

- dairy cattle that produce lots of milk
- tomato plants that produce a high yield of tomatoes
- wheat that is resistant to a particular pest.

Selective breeding is one of the tools agricultural scientists have been using to try to produce enough food for everyone. This is becoming increasingly important as the world's population increases and people are living longer.

> **a** What is meant by the term 'selective breeding'?

## Selectively breeding sheep

Sheep farmers will select characteristics based on the produce they sell. Farmers who breed sheep for wool choose sheep that produce large, good quality fleeces. These can then be manufactured into wool products such as yarn or carpets. Farmers that raise sheep for meat, choose breeds and individuals that produce lean meat and a large number of offspring. Some farmers also raise sheep to produce cheese; they select individuals that have a high milk yield.

> **b** How does the farmer choose which of the offspring to breed from?
>
> **c** Why does it take several years of selective breeding to produce a flock of sheep with the desired characteristics?

### Activity

**Pedigree dogs**

Pedigree dogs are selectively bred. They have the desired characteristics of their breed. However, many suffer from health problems as a result.

For example, the desire to have a sloping back with hind legs low to the ground causes hip problems in German shepherds, and the preference for a snub nose causes breathing problems in pugs. This is made worse when animals are 'in-bred'. This means closely related dogs, such as siblings, are mated. As a result of in-breeding, pedigree dogs have a much lower life expectancy than cross-breeds.

- Choose a species of pedigree dog.
- Research the desired characteristics of the breed, and any related health problems from which they suffer.
- Present your information as a leaflet or a poster that could be given to people researching which breed to choose as a new family pet or working dog.

## Selectively breeding wheat

Wheat is a very important crop in the agricultural industry. It produces grain that is turned into flour, which has a wide number of uses, such as making bread. Selective breeding by farmers has changed the characteristics of wheat, as the table and Figure 2 show:

| Wild wheat plants | Modern wheat plants |
|---|---|
| Ears are small and have few seeds | Ears are large and have many seeds |
| Stalks are brittle and ears often fall off | Stalks are stronger, so ears stay on |
| Ears ripen at different times | Ears ripen at the same time |
| Stalks grow to different heights | Stalks grow to the same height |

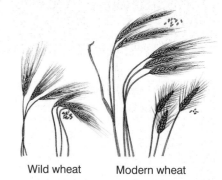

Wild wheat        Modern wheat

**Figure 2** Effect of selective breeding of wheat

**d** How do the features of modern wheat make it easier to harvest?

Many modern crops have also been selectively bred to have a high resistance to disease.

## Problems of selective breeding

Selective breeding reduces the number of genes (the **gene pool**) from which a particular strain or variety of a species is created. It reduces variation. This means that, if a new disease arises, it might be that none of the organisms in that gene pool have the gene for resistance to this disease. This could result in a particular strain becoming extinct.

### Summary questions

1 Copy and complete the table with the characteristics a farmer may choose to breed selectively for:

| Organism | Characteristics |
|---|---|
| Sheep | |
| Chicken | |
| Cow | |
| Wheat | |
| Apple tree | |

2 What are the disadvantages of selective breeding?

3 A farmer wants to grow large, tasty strawberries. One species of strawberry is large and tasteless. Another species produces very small but very sweet and juicy strawberries. Draw a diagram to show how a farmer can selectively breed for the large and tasty characteristics.

### Key points

- Selective breeding involves choosing the best organisms to breed to produce offspring with desired characteristics.

- Farmers selectively breed sheep for wool by breeding ewes and rams with the longest fleeces.

- Selectively bred wheat has larger ears with more seeds. The stalks are strong and grow to a similar height and the grain ripens at the same time. This makes the crop easier to harvest and produces higher yields.

# 13.2 Genetically modified food

## Learning objectives

- What is genetic engineering?
- How are plants and animals genetically modified?
- What are some examples of genetically modified plants?

**Figure 1** Genetically modified cotton has high yield and pest resistance

**Figure 2** The glo-fish, which is genetically engineered to fluoresce. Scientists are trying to develop the fish so that it only glows when it comes into contact with environmental pollution.

**Figure 3** Genetically modified tomatoes can be made frost resistant by adding genes that code for an antifreeze chemical to be produced

## Genetic engineering

Farmers use selective breeding to produce animals with desired characteristics. This is a slow process and the outcome is not very predictable. In the past, farmers did not realise that, during this process, they were changing combinations of an organism's genes.

Scientists now have a much greater understanding of genetics. They are able to alter the combinations of genes in an organism to produce the desired characteristics. This is called **genetic engineering** (or **genetic modification**). It can happen in one generation.

**a** What is genetic engineering?

**b** Name **two** advantages of genetic engineering over selective breeding.

### Activity

#### Genetically modified crops could feed the world

Genetically modified (GM) crops were first sold in 1996. However, how safe they are and whether they damage the environment are still being debated.

Each year the world's population increases by more than 80 million people. To feed this number of people by traditional farming methods, all usable soils in the world would have to be cultivated. This would destroy many habitats. Crops can be genetically modified to give a higher yield than a traditional variety. The ultimate answer to this is population control, but in the meantime, GM crops *could* help to feed the world without destroying large areas of the environment.

However, many people are concerned that genetically engineered crops may interact with plants in the surrounding environment. The long-term effects of cultivating GM crops are currently not known.

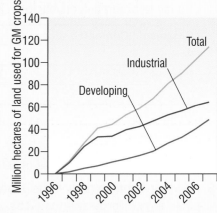

1 Look at the graph above and answer the following questions.
   **a** Why are more GM crops grown in industrialised countries?
   **b** Why is the growth of GM farming slower in the industrialised world than the developing world?

2 If tropical fruits like bananas could be genetically engineered to live in cool climates, how would this harm the economy of poorer countries?

3 Most people prefer seedless grapes. Other fruits could be genetically engineered not to contain seeds. These plants could cross-pollinate with wild plants. What effect could this have on the native plants in an environment?

## How does genetic modification work?

Genes from another organism (foreign genes) are put into plant or animal cells at a very early stage in their development. As the organism develops, it will display the characteristics of the foreign genes.

**c** Why are the genes inserted into the plant or animal cells known as foreign genes?

## Making frost-resistant plants

Bt corn (also known as transgenic maize) is an example of a genetically modified crop. Scientists isolated a gene from *Bacillus thuringiensis* (a type of soil bacterium) and put it into corn. This gene codes for a toxin that kills insects. This means that insecticides are not needed for this crop.

The flow diagram below shows how crops are genetically modified.

A useful gene is removed from the nucleus of a donor cell.

⬇

The foreign gene is then put into a circular piece of DNA called a **plasmid**. This is now known as a piece of **recombinant DNA**.

⬇

The recombinant DNA is put into a bacterial cell.

⬇

Plant cells are infected with the bacteria. The foreign gene becomes integrated with the DNA of the plant cells.

⬇

The plants cells are placed in a growing medium to grow into plants. These plants will have the desired characteristics.

**d** Name an advantage of growing Bt corn over traditional varieties.

Animals are genetically modified in a similar way, by inserting the required genes into an embryo. This technique has been used to produce a range of genetically modified animals, such as laboratory mice and rats. The use of animals in this type of research is strictly controlled.

### Summary questions

1 Copy and complete using these words:

*chemicals foreign resistant characteristics genetic*

It is possible to insert ............ genes into organisms, to change their ............ . This is called ............ engineering. Plants can be produced that are ............ to disease. Microorganisms can be engineered to produce useful ............ .

2 Name **two** examples of genetically modified plants and animals.

3 Why must the foreign genes used in genetic engineering be inserted into an embryo and not into a more developed organism?

### Key points

- Genetic engineering involves changing an organism's genes, so they produce desired characteristics.

- Genetically modified organisms are produced by putting foreign genes into plant or animal cells at a very early stage in their development. As the organism develops, it will display the characteristics of the foreign genes.

- Genetically modified plants include frost-resistant tomatoes, high-yield cotton and insect-resistant Bt corn.

<table>
<tr><td>**13.3**</td><td># Genetic engineering and cloning</td></tr>
</table>

## Learning objectives

- How can bacteria be genetically engineered?
- What is a clone?
- What are the differences between genetic engineering and cloning?

**Figure 1** *E. coli* is genetically engineered to produce insulin

**Figure 2** Bacteria are placed in a fermenter. This provides the perfect conditions for bacterial replication: warmth and nutrients.

### ??? Did you know …?

Many plants produce clones of themselves naturally. For example, strawberries send out runners, from which a new genetically identical plant will grow. This is known as asexual reproduction – meaning from one parent.

## Genetic engineering of bacteria

Bacteria can be genetically engineered to produce many useful chemicals. These include hormones, vaccines and antibiotics.

Genes that code for the production of the required chemical are inserted into the bacteria. The bacteria reproduce, producing millions of identical copies of the gene, and therefore the substance it codes for. The useful substances are then separated from the bacteria.

> **a** Name **three** medical substances that can be produced by genetically engineered bacteria.

## Insulin production

Bacteria have been genetically engineered to produce **insulin**. This hormone is used to control the blood glucose levels of many people suffering from **diabetes**. As bacteria reproduce very quickly, large amounts of insulin can be produced in a short period of time.

Bacteria are made to produce insulin by the following process:

1 Small circles of DNA called **plasmids** are found in bacteria. These are modified to include the piece of human DNA containing instructions for making insulin.
2 The plasmid is then inserted into the bacteria *Escherichia coli*. The bacteria now produce insulin.
3 The bacteria multiply, and produce large quantities of insulin in a fermenter.
4 Bacteria are then killed by heat (sterilisation), leaving behind the insulin.

> **b** Why do bacteria reproduce better in a fermenter?

## Cloning

When microorganisms reproduce, they produce genetically identical copies of themselves. These are called **clones**. This is exploited when bacteria are used to make useful chemicals.

> **c** What is a clone?

## Plant cloning

Taking a plant cutting is an example of cloning. Gardeners have been using this technique for centuries to produce more plants. Biotechnologists have developed this technique further, to help grow more food. Plant tissue culture allows the rapid production of many genetically identical plants. This is also known as **micropropagation** (see Figure 3).

Micropropagation offers several advantages over traditional planting techniques. Large numbers of plants can be produced, very quickly and need very little space. The plants are in a controlled environment and so are disease free.

The main disadvantage of cloning plants is in the reduction of the gene pool. This can increase the risk of disease destroying a species.

## Animal cloning

Human cells can be cloned in the laboratory and used for research into diseases and new drugs. For example, tissue culture is used to produce skin and cartilage.

This is carried out in the following way:

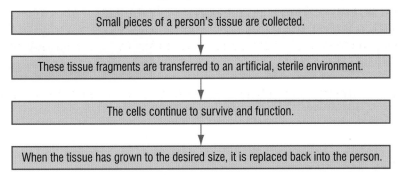

| Small pieces of a person's tissue are collected. |
| --- |
| These tissue fragments are transferred to an artificial, sterile environment. |
| The cells continue to survive and function. |
| When the tissue has grown to the desired size, it is replaced back into the person. |

Tissue culture is used in skin grafting. New skin tissue is produced, which can be used to repair damage caused by burns or serious injury. As the new tissue is genetically identical to the donor material, the body will not recognise it as being 'foreign'. This minimises the risk of rejection.

**Differences between genetic engineering and cloning**

|  | Genetic engineering | Cloning |
| --- | --- | --- |
| Genes present in organism | Produces a new set of genes – different from the original | Produces an exact copy of the genes present |
| Origin of genes | Uses 'foreign' genes, often transplanted from a different species | Uses genes copied from a member of the same species |
| Characteristics produced | Gives a new, desired characteristic to the organism | Produces a copy of the original characteristics |

## ⚭ links

*For more information on auxins (plant hormones) and their role in plant growth look back at 3.6 Plant growth.*

1. A tissue sample is scraped from the parent plant (only a few cells are needed).

2. Tissue samples are placed on an agar plate containing nutrients and auxins. Auxins are plant growth hormones.

3. Samples grow into tiny plants (plantlets).

4. Plantlets are planted in compost, and grown in a greenhouse where they develop into full-sized plants.

**Figure 3** The process of micropropagation

## Summary questions

1 Match the following to make three sentences:

| Cloned organisms | is a technique used to rapidly produce many identical plants |
| --- | --- |
| Micropropagation | produce many medicines including insulin |
| Genetically engineered bacteria | are genetically identical to their parent |

2 Name **one** advantage and **one** disadvantage of cloning plants.

3 Draw a series of labelled diagrams to show how tissue culture can be used to produce many blueberry plants.

### Key points

● Genes can be inserted into bacteria to make them produce useful chemicals. These include insulin and vaccines.

● A clone is a genetically identical copy of an organism.

● Genetic engineering alters the genes of an organism; cloning produces exact copies of an organism.

## 13.4

# Ethics of genetic engineering and cloning

## Genes and ethics

*Science Weekly*

### 'Genetic engineering to boost human intelligence'

Scientists have revealed that a gene has been identified, NR2B, which could help to boost human intelligence. The gene, which was inserted into laboratory mice, has helped the mice to learn and remember. Research scientist Dylan Gray said simply, 'It just makes them smarter'.

The research could one day lead to an increase in the brainpower of human beings, or to a cure for degenerative diseases of the brain, such as dementia. However, campaign group 'No to GMO' has highlighted the potential dangers of the research, including not knowing the long-term effects of the gene therapy.

**a** Read the article above. Name **one** potential benefit and **one** drawback of research into NR2B.

There is a lot of controversy surrounding research into genetic engineering, cloning and into gene-replacement therapy. Supporters of the research say that scientists have the potential to tackle huge issues. These include feeding the world's increasing population and providing cures and treatments for a number of diseases. Others believe these techniques are unethical and should be banned. Concerns have been raised over how individual organisms will be affected and the long-term effect on the ecosystems in which these organisms live.

**b** Using Figure 1, draw a table that summarises **two** arguments for, and **two** arguments against genetic research.

### 'Designer babies'

Geneticists help couples with fertility problems to have babies. One technique used is *in vitro* fertilisation (IVF). Human eggs are taken from the woman and fertilised with sperm, in the laboratory. The embryos are then allowed to develop for a few days, before being implanted in the woman's uterus.

Some couples also choose to have their babies through IVF if there is a risk they might have an inherited disease. The embryos created can be screened for genetic disorders. Only healthy embryos are selected to be implanted.

There are ethical issues with IVF. People are concerned that couples may want to choose 'designer babies' with desirable qualities – such as choosing to have a girl if the couple already have boys.

**c** Using the information above, name one advantage and one disadvantage of IVF.

Organisms can be genetically modified to produce medicines very cheaply. For example, sheep could be used to produce proteins that help to relieve the symptoms of cystic fibrosis.

Plants can be modified to glow in the dark when they need watering. This will reduce how much water farmers use – which is really important in countries that do not have much water available.

Seedless fruits have been developed. Although many people prefer eating these, no one knows what may happen if these species interact with wild species. We could end up with no seed-producing plants.

Altering an animal's genes affects the way the organism functions. When an organism is genetically engineered, no one really knows whether or not the animal will be in pain, or have a shorter lifespan. All living organisms have a right to quality of life.

**Figure 1** People have different viewpoints on genetic research. What do you think? There is not necessarily a right or wrong answer to this issue.

## Activity

### Would you like to be cloned?

Animal cloning has many important commercial implications. It allows an individual animal with desirable features, such as a cow that produces a lot of milk, to be duplicated. The process takes much longer than it does with plants.

● Dolly was the first mammal to be cloned. She died prematurely.
● Only 700 mountain gorillas are left. Scientists could clone more.
● People's bodies would not reject cloned copies of their own organs. These could be used to replace their damaged or diseased organs. However, scientists can't 'make' cloned organs yet.

1 Do you agree with animal and human cloning? Make a list of all the possible advantages and disadvantages.
2 Should there be strict guidelines for cloning? Discuss your views in your group.

## Gene replacement therapy

Some people are born with genetic defects. These sometimes lead to a person suffering from a genetically inherited disorder, such as Polydactyly. These disorders are impossible to cure using traditional medical techniques.

Gene replacement therapy is the replacement of the faulty gene with a healthy gene. This would lead to the disease being cured.

Although many people support this research, others are concerned about the possibility of gene replacement being used for non-medical purposes. For example, to improve your appearance, or to make you stronger or faster.

## The ethics of genetic engineering and selective breeding

Selective breeding has been used by farmers for thousands of years. It uses a natural approach – the combination of genes from each of the parent organisms.

Genetic engineering has been developed in the past 20 years. It is not natural and relies upon gene manipulation in the laboratory – often crossing genes between species.

Many people have few concerns over selective breeding, but do not approve of genetic engineering. However, both processes are essentially the same. They involve the human selection of genes to produce a desired organism.

### AQA Examiner's tip

The ethics of genetics is a hot topic of debate. There is no right or wrong answer. To succeed in an exam you should understand people's different points of view, and be able to justify your own beliefs using scientific reasons

### Key points

● Genetic research is highly controversial as the techniques involve altering the genetic material present in an organism.

● A 'designer baby' develops from an embryo that has been created specifically for a desired characteristic.

● In gene replacement therapy a faulty gene will be replaced with a healthy gene. This technique has the potential to cure many genetically inherited disorders.

## Summary questions

1 Copy and complete using these words:

   *engineering   natural   genetics   animals*

   Research into ............. is controversial. Some people think genetic .............
   is not a ............. process, and so should not happen. Others believe
   that it could help to produce crops and ............. that will feed the world's
   growing population.

2 a What is the name of the laboratory process which can be used to help couples to conceive a child?
   b What is meant by the term 'designer baby'?

3 Do you think genetic engineering should be allowed? Write a reasoned argument for, or against, the continuation of this research.

# Summary questions

1 The following sentences explain how farmers may selectively breed their crops. Rearrange them into the correct order:
- Offspring grow.
- Farmer chooses crops with the best characteristics.
- Farmer then chooses crops with the best characteristics.
- These crops are then cross-pollinated.
- These are then bred. This process is repeated for many years.

2 The Flvr Svr tomato is genetically engineered to stay firm for longer.
a Why is this an advantage for tomato sellers?
b How are the genes in this tomato's nucleus different from those of a normal tomato?
c Why is this process quicker than selective breeding?
d What other features might a tomato grower choose to improve?

3 Some people require an organ transplant, due to damage or disease. Animals could be genetically modified, so they can provide organs for humans.
a Where do organs for donation currently come from?
b What are the arguments in favour of using animals in this way?
c What are the arguments against using animals in this way?

4 Bacteria can be genetically engineered to produce insulin.
a What is a plasmid?
b What is a clone?
c Name two advantages of producing insulin using bacteria.

5 a What is the technique of micropropagation used for?
b Name two advantages of this technique
c Name one disadvantage of this technique.

6 The following sentences describe how a plant can be genetically modified to display a specific characteristic. Rearrange the sentences into the correct order:
- Plant cells are infected with the bacteria.
- The foreign gene becomes integrated into the plant's DNA.
- A useful gene is removed from a donor nucleus.
- The bacteria reproduce producing lots of copies of the recombinant DNA.
- The recombinant DNA is put into a bacterial cell.
- The foreign gene is put into a plasmid.
- The plants cells are grown into plants that display the desired characteristics.

# AQA Examination-style questions

**1** Genetic engineers have been working to improve organisms. The table shows possible outcomes of genetic modification.

Tick the boxes to show which of these are advantages, disadvantages or neither.

| Outcome | Advantage | Disadvantage | Neither |
|---|---|---|---|
| Could spread and destroy native species | | | |
| Glow in the dark pigs | | | |
| Herbicide-resistant wheat | | | |
| Possible introduction of cancer | | | |
| Removal of genetic disorders | | | |

(5)

**2** Farmers have been using selective breeding for centuries.
**a** Put the sentences in the correct order to describe the process of selective breeding of sheep.
**A** Farmer repeats over several generations
**B** Farmer chooses the best offspring
**C** Farmer chooses sheep that have lots of wool
**D** Farmer breeds the sheep (3)
**b** Suggest another quality of sheep that a farmer might want to improve. (1)
**c** Give a disadvantage of selectively breeding sheep. (1)

**3 a** *In this question you will be assessed on using good English, organising information clearly and using specialist terms where appropriate.*
Human insulin can be made using genetically modified bacteria. Explain how they have done this. (6)
**b** Before human insulin was produced through genetic engineering, diabetics used to be treated with insulin from pigs. Suggest **one** disadvantage of this treatment.(1)

**4** Greyhound dogs have been selectively bred to run very fast in races. The table shows the average time for a greyhound to run 2 km in different years.

| Year | 1920 | 1930 | 1940 | 1950 | 1960 |
|---|---|---|---|---|---|
| Time in seconds | 161.00 | 160.02 | 157.88 | 152.25 | 149.86 |
| Year | 1970 | 1980 | 1990 | 2000 | |
| Time in seconds | 145.32 | 142.35 | 138.26 | 136.85 | |

**a** Suggest the average time to run 2 km in 1935. (1)
**b** In which year might it have taken 139.28 seconds? (1)
**c** Describe an experiment that greyhound racers could do to collect results for this year. (3)
**d** Calculate the percentage decrease in the overall time taken to run 2 km. (4)

**5** A couple who have a son with a blood disease researched different treatments. They read this newspaper article on 'designer babies'.

## Designer babies could save money for the NHS

Children who are very ill with incurable diseases might require a lot of medication for long periods of time. That could cost the NHS as much as £50,000 a year and nearly £2 million in a lifetime.

It has been estimated that the cost of producing a designer baby through IVF treatment is £20,000.

Dr Edwards explained how a designer baby is chosen. 'Tests are conducted on the mother's embryos to find one that does not have their brother or sister's inherited disease and is a suitable genetic match. It is then implanted in the mother's womb to create a baby from which bone marrow or cells can be transplanted into their ill sibling.'

**a** Suggest why the couple might be against having a designer baby. (1)
**b** If the couple's son lived until he was 20 years old, what could be the saving for the NHS if he could be treated using a 'designer' sibling? (4)
**c** What could be the long-term implications of being able to select the characteristics of a baby before it is born? (2)

**6** Selective breeding and genetic modification are both methods used to change the characteristics of organisms.

Sort these advantages out into those about selective breeding, those about genetic modification and those that apply to both.
• Bigger crops
• Get desired characteristics
• Easy for farmer to do
• Quick

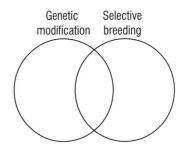

(2)

# Improving our environment

In Unit 3, Theme 3 you will work in the following contexts, covered in Chapters 14 and 15:

## Environmental concerns when making and using products

Manufacturing new products is not always good for the environment. The more materials are processed, the more energy is used to produce them. Generating this energy usually produces greenhouse gases, which are contributing to climate change. Other types of production can also harm the environment. Growing enough food to feed everyone requires the use of fertilisers. These can be washed into rivers and end up killing fish through a process called eutrophication.

Getting rid of old products can be as damaging as making new ones. Landfill sites all over the country are filling up as we try to find new ways to dispose of waste. To help with this problem, materials scientists are constantly trying to invent new forms of biodegradable packaging.

## Saving energy in the home

### Heat transfers

The kitchen is a room of extremes of temperature. Fresh food is stored at low temperatures in fridges or frozen at sub-zero temperatures in freezers. We boil water in kettles and cook food at much hotter temperatures. How can we do this as efficiently as possible in the same room? In different buildings, temperatures are controlled by controlling heat transfers so that the temperature inside is comfortable whether there is ice outside or blazing sunshine.

### Saving money on our heating bills

Heating bills can be very expensive for some houses, but neighbouring houses may have very different heating bills. Modern homes are built to be well insulated due to our building regulations but if you have an older house what can you do to reduce your bills? Some owners spend large amounts of money on changes that are not as cost-effective as a quicker simpler change. How can you tell if a change is worth doing? Householders need to know if the amount of energy saved is significant and how big the saving will be.

## Controlling pollution in the home

### Pollution in the home

We are used to pollution being a problem caused by traffic and power stations. It is surprising to realise there can be pollution inside our homes as well. Dust, moulds, spores and pollen can all build up inside homes to levels that cause ill effects in people.

### Household hazards

We use many products in our homes to keep them clean and free from germs. Some products have been used for years but other newer products are less familiar. The packaging carries a lot of (hazard) information about how to store or use a product safely.

### The silent killer

Homes use boilers to heat water and for central heating. Most of the time, we take them for granted until something goes wrong. A faulty boiler may just mean a cold shower in the morning but in some cases faulty boilers can result in deaths.

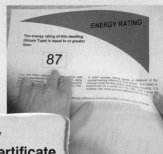

## Faulty gas boiler causes woman's death

A woman was killed and her husband is fighting for his life after suffering from carbon monoxide poisoning. This is thought to be a result of a faulty boiler installed in the home. The victim would not have realised she was being poisoned as the gas has no smell, taste or colour. Carbon monoxide monitors would have prevented this death.

## Was the Energy Performance Certificate worth having?

One man is angry after he bought a flat, which was rated as energy efficient, and likely to have low heating bills. After a year in the flat, his heating bills mounted up to thousands of pounds and the flat always felt cold and draughty. He thinks he was mis-sold the flat and is demanding compensation from the provider of the Energy Performance Certificate for the high heating bills.

## Sick building syndrome closes offices

Tightly sealed windows and doors save energy and money, but can they make workers ill too?

One new office building was so effective at sealing in the air, workers became ill with a range of symptoms ranging from headaches, tiredness, coughs and sore throats. House plants have now been introduced to these offices, and some windows unsealed so that the workers carry on working feeling as fresh at the end of the day as when they arrived.

## Landfill leakage causes loss of life

Hundreds of fish and birds were killed by toxic liquids seeping out of a UK landfill. Experts believe the thick lining of the landfill had become damaged and allowed the poisonous chemicals out. Local residents are now campaigning to have the landfill closed. They want a modern waste incinerator installed instead.

## 14.1 Producing greenhouse gases

**Figure 1** Burning fossil fuels produces carbon dioxide and other greenhouse gases

We use many different products in our daily lives compared with earlier generations. The quantity of different products has increased dramatically. We buy goods more often and throw more away. More people live on Earth now than ever before. This all adds up to a big impact on the environment worldwide. Scientists are now looking for more environmentally friendly ways to:

- manufacture products
- reduce their impact when in use
- reduce their effect on the environment when they are thrown away.

Scientists are very concerned about three gases in particular: **carbon dioxide**, **methane** and **nitrous oxide**. These gases are produced in large quantities and are examples of **greenhouse gases**. Our actions have increased the concentration of these gases found in our atmosphere. The atmosphere traps more of the Sun's energy when the concentration of greenhouse gases increases. This effect is called **global warming** and it is expected to change our climate in future. There is some evidence to suggest this is already taking place.

a What are the main greenhouse gases?

b What is global warming?

### Carbon dioxide

Carbon dioxide is the greenhouse gas produced in largest quantities. The quantity produced when **fuels** are burned has increased greatly in the last 70 years. The **fossil fuels** – coal, oil and gas – have taken millions of years to form. Fossil fuels are burned in increasing amounts in vehicles and in power stations to generate electricity. Fossil fuels are also used for heating and cooking. Smaller quantities of fuels such as wood and **biofuels** are also burned, and they produce carbon dioxide too.

c How does burning fossil fuels contribute to global warming?

### Methane

Methane is 20 times more effective at trapping heat than carbon dioxide per molecule. It is the next most important overall contributor to global warming after carbon dioxide. Methane is released into the atmosphere as part of the carbon cycle in several ways:

- from the action of bacteria when plants and other organic materials decay with very little air present – this can happen in swamps, rice paddy fields and hydroelectric schemes
- when organic material rots in a landfill site or compost heap
- from coal mining activities and other industrial processes
- from bacteria in the guts of animals, including farm animals.

**Figure 2** Bacteria in paddy fields release methane, a greenhouse gas

d Write down **four** ways that methane can be produced.

## Activity

### Greenhouse gases from different countries

Find out how the amount of each of the three main greenhouse gases produced by different countries varies.

Present your data to the class either as a poster or PowerPoint presentation.

### Nitrous oxide

Nitrous oxide is about 310 times more effective at absorbing energy than carbon dioxide per molecule. It is produced naturally, but increasing amounts are produced from agriculture.

Intensive farming methods use nitrogen-based fertilisers to increase crop yields. If large amounts of fertiliser are used, denitrification can take place. **Denitrification** is when bacteria convert nitrates into nitrous oxide if there is little oxygen present in the soil. Using animal waste in large quantities as a fertiliser also produces nitrous oxide.

Nitrous oxide is also made when fuels are burned at very high temperatures, as in car engines. Nitrogen is a relatively unreactive gas, but in these conditions it does react with oxygen. However, catalytic converters in cars can convert nitrous oxides in exhaust gases into harmless nitrogen gas.

 **links**

*For more information on the effects of greenhouse gases see 14.2 The effects of greenhouse gases.*

**Figure 3** Intensive farming contributes to global warming

**Did you know …?**

You can calculate how much your lifestyle and actions contribute to the main greenhouse gases – this is called your carbon footprint. Carbon footprint calculators can be found online.

e  Write down **two** ways nitrous oxides are produced in large quantities.

## Summary questions

1 Here are three of the main greenhouse gases:

**carbon dioxide   methane   nitrous oxide**

a Put the gases in order of the quantities found in the atmosphere, starting with the most abundant.

b Put the gases in order of their 'warming effect', starting with the gas that has the biggest 'warming effect' per molecule.

2 Explain why putting waste on a compost heap can produce the same greenhouse gases as sending it to a landfill site.

3 Prepare a leaflet summarising the main ways that each of the three greenhouse gases in question 1 are produced.

### Key points

- Greenhouse gases include carbon dioxide, methane and nitrous oxide.
- Carbon dioxide is produced when fuels are burned.
- Methane is mainly produced when bacteria break down organic matter in the absence of oxygen.
- Nitrous oxide is produced by nitrogen-based fertilisers from intensive farming and by car engines.

# 14.2 The effects of greenhouse gases

## Learning objectives

- Why does global warming occur?
- What is the Kyoto protocol?
- Why do we need the Kyoto protocol?

The Sun is so hot it radiates short wavelength light and ultraviolet radiation. Our **atmosphere** reflects some of the Sun's radiation, allowing about two-thirds to pass through. The Earth absorbs this energy, and warms up. As Earth warms, it emits more radiation as long wavelength (infrared) radiation. If the amount of energy absorbed matches the energy emitted, the temperature of Earth stays constant.

**a** What happens to the Sun's radiation when it reaches our atmosphere?

Some gases in the atmosphere are effective at absorbing infrared radiation radiated by the Earth. They re-radiate it in all directions so some is directed back to Earth. This raises the temperature in our atmosphere higher than it would be otherwise. This is called the **greenhouse effect** and is one reason why the Earth can sustain life. Gases like carbon dioxide, methane and nitrous oxides are called **greenhouse gases.**

The balance of gases in our atmosphere has changed over the last century. Burning fossil fuels releases carbon that was trapped in the Earth to the atmosphere as carbon dioxide. The proportion of gases like carbon dioxide, methane and nitrous oxides has increased in our atmosphere. Gradually our atmosphere is warming up. This increase in the average temperature of the Earth's atmosphere is called **global warming**.

**b** What do greenhouse gases do to infrared radiation?

**c** What is meant by global warming?

**Figure 1** Global warming occurs when more radiation is absorbed than is emitted

## Global warming

Global warming is worrying because scientists believe it is changing our climate. No one knows how much the temperatures will rise, or what the effects will be. Some effects could be:

- there may be more droughts
- sea levels may rise
- polar ice caps and glaciers may melt
- there may be more hurricanes and storms.

These effects are called **climate change** and could have a big impact on our way of life.

**d** What are the main effects of climate change?

One country on its own cannot control global warming. The UK only contributes about 2 per cent of the world's greenhouse gases. Gases from all countries spread throughout the atmosphere. If a country is producing less greenhouse gases, it may be less able to trade with other countries. That is because it is manufacturing fewer products or having to raise costs as a result of controlling emissions. If all countries agree to reduce activities producing greenhouse gases, we may be able to limit climate change.

## The Kyoto agreement

The **United Nations** is a group of industrialised nations including the European Union. In 1997 many of its representatives met in Kyoto, Japan. Some of them signed an international agreement called the **Kyoto Protocol**. It set targets for these countries to reduce their greenhouse gas emissions by about 5 per cent of their 1990 levels over the years 2008–12. It was a legally binding document, forcing them to monitor their emissions, meet the targets and report back. This is an important first step, but some countries did not sign and others did not get the Kyoto protocol ratified (approved) by their governments.

⊂⊃ **links**

*For information on producing greenhouse gases look back at 14.1 Producing greenhouse gases.*

- ☐ Ratified
- ■ Signed only
- ■ Not signatory
- ☐ No data

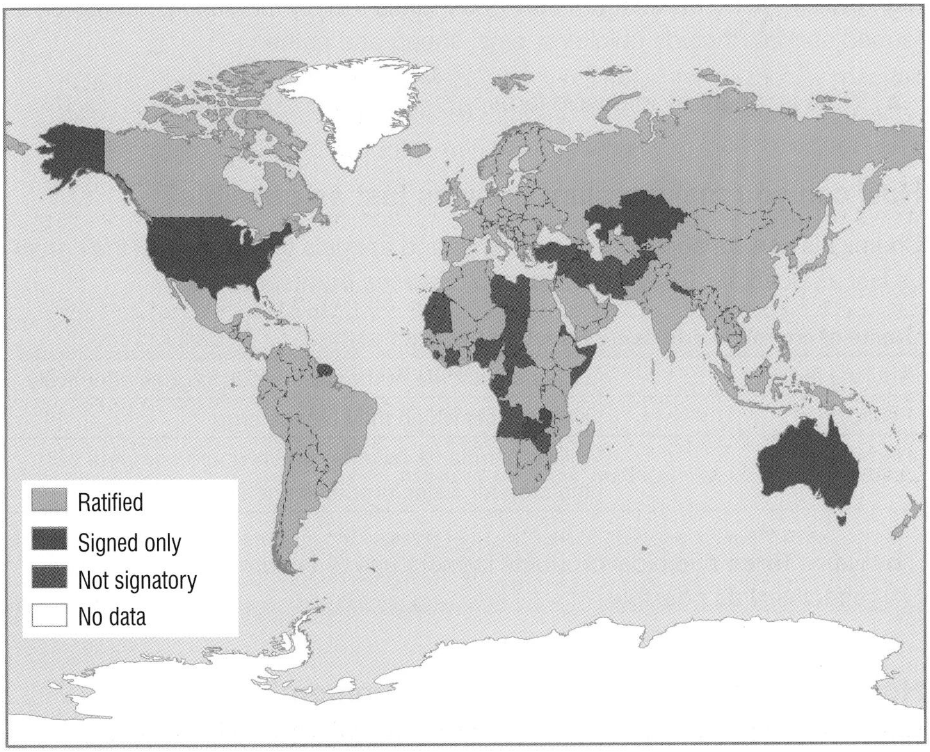

**Figure 2** Many countries signed the Kyoto Protocol but some did not

**?? ?** **Did you know …?**

The UK is reducing its emissions by making homes more energy efficient, producing more energy from renewable sources, setting limits on business emissions and raising taxes on landfill sites to encourage recycling.

**e** What was the purpose of the Kyoto Protocol?

### Summary questions

1 Copy and complete using these words:

*global warming   shorter wavelength   longer wavelength
climate change*

The Sun radiates ............ radiation and the Earth emits ............ radiation.

............ is when temperatures in our atmosphere increase.

............ may cause more droughts, hurricanes and rising sea levels.

2 Write a sentence that explains the difference between greenhouse effect, global warming and climate change.

3 Explain why one country cannot tackle climate change on its own.

4 Find out how the Kyoto Protocol has affected different countries in the last ten years.

### Key points

- The greenhouse effect is when energy is retained in the atmosphere by greenhouse gases.

- Global warming is the increase in the average temperature of our atmosphere.

- Global warming will cause climate change.

- The Kyoto Protocol is a legally binding agreement.

- Many countries in the world agreed to cut their emissions of greenhouse gases.

## 14.3

# Threats to the countryside

### Learning objectives

- Which chemicals are used in intensive farming?
- What is eutrophication?
- What are indicator species?

**Intensive farming** is farming that produces as much food as possible in the space available. This is achieved by making the best use of land, plants and animals. It makes the food produced as cheap as possible.

Intensively farmed animals are kept in a strictly controlled environment. The conditions include keeping the animals warm, and restricting their space. These aim to prevent their energy being wasted. The animals are also fed a high protein diet. This ensures their body mass rapidly increases. Intensively farmed animals include chickens, pigs, sheep and cattle.

**a** What is meant by intensive farming?

### How can you make a plant grow as fast as possible?

Chemicals can be applied to both plants and animals to ensure that they grow as fast as possible. They can also stop diseases from spreading.

| Name of chemical additive | Effect on the crop |
|---|---|
| Artificial fertiliser | Gives a plant the nutrients it needs to grow effectively |
| Pesticide | Kills insects which may eat the crop |
| Herbicide | Kills other plants (weeds) which would compete with the crop for water, nutrients and space |

**b** Name **three** chemical products farmers use to ensure crops grow as effectively as possible.

### How can fertilisers damage the environment?

NPK is a common fertiliser. It contains three of the essential minerals that plants need for healthy growth, nitrogen (N), phosphorus (P) and potassium (K). If farmers use large amounts of fertiliser, it can wash off the land when it rains. This can drain into water sources. This leads to **eutrophication**. You need to know the steps involved, shown in Figure 1.

**c** What causes eutrophication?

### How do scientists monitor pollution?

Scientists regularly take samples of plants and animals from the environment, to monitor the type and number of organisms present. If the number or range of species decreases this could indicate that there is pollution.

**Indicator species** are species of organisms that can be used to measure environmental quality. The numbers of indicator species increase or decrease in the presence of certain levels of pollutants.

**Fertiliser run-off**

1. Excess fertiliser dissolves in water in the soil and runs off fields into rivers and lakes. This process is known as leaching.

2. The fertiliser causes water plants such as algae to grow rapidly. The algae quickly cover the surface of the water, stopping light reaching lower plants.

3. The dead plants and algae are broken down by microorganisms. The process of decay uses up lots of oxygen dissolved in the water. This makes it increasingly difficult for animals to survive, and many fish die.

**Figure 1** The main steps in eutrophication

### AQA Examiner's tip

You may be asked to describe all the steps involved in eutrophication in the exam. Draw a flowchart to help you remember.

## Monitoring air pollution

Lichens are often used to monitor air pollution. As they have no root systems, most of their nutrition comes from the air. Many species cannot live in areas with high concentrations of acidic sulfur dioxide gas. For example, lichen cannot survive near factories that produce large quantities of toxic gases, such as metal smelters.

## Monitoring water pollution

Aquatic invertebrates and fish can be used as indicator species of water quality. If only sludge worms are present in a water source, it may indicate sewage contamination. Sludge worms can live in water containing virtually no oxygen. Most animals cannot.

Large increases in the numbers of sludge worms can be found immediately downstream of a sewage leak. Further upstream, sludge worm numbers will be low, reflecting the cleaner conditions. Further downstream from the leak, as the discharge becomes more diluted, the number of worms decreases.

Mayfly larvae are an indication of clean water. In a river receiving waste water from a sewage treatment plant, mayfly larvae will be present in large numbers before the discharge point. Their numbers will dramatically decrease at the discharge point. They will increase again further downstream, as the effects of the sewage discharge are diluted.

| POLLUTION LEVEL INDICATOR SPECIES | | |
|---|---|---|
| **Pollution level** | **Examples of organisms present in water** | |
| Clean | Stonefly larva | Mayfly larva |
| Low | Freshwater shrimp | Caddis fly larva |
| High | Bloodworm | Water louse |
| Very high | Rat-tailed maggot | Sludge worm |

**d** How can the level of air pollution be shown by lichen species?

**e** Name **two** indicator species of water pollution.

**Figure 2** *Usnea subfloridana* is a species of lichen that cannot survive in high sulfur dioxide concentrations

### Did you know … ?

Canaries are more sensitive to carbon monoxide than humans. They were used by miners to detect gas. If a canary died underground, the miners knew that they must immediately evacuate the mine. This is one of the first known uses of indicator species.

**Figure 3** These are sludge worms. If they are the only species present in water, then this indicates that the water is contaminated with sewage.

### Summary questions

1 Copy and complete using the words below:

*eutrophication chemicals fertilisers controlled species intensive pollution*

In ............. farming, farmers make as much food as possible from the land available. This involves adding ............. to crops, and keeping animals in a ............. environment. If ............. run off into lakes, they can cause .............

Indicator ............. are organisms which can be used to detect .............

2 What happens if fertiliser levels build up in a lake?

3 If a lake became polluted with sewage, describe the numbers of mayfly larvae and sludge worms that would be found:

   **a** upstream of the leak.

   **b** at the leak.

   **c** a few miles downstream of the leak.

### Key points

- Fertilisers, pesticides and herbicides are used to increase crop production in intensive farming.

- If fertiliser drains into rivers and lakes it can lead to eutrophication.

- Indicator species monitor the presence of pollution. Their numbers increase or decrease in the presence of certain pollutants.

## 14.4

## Disposing of our waste

### Learning objectives

- How do we dispose of our waste?

- How can waste be disposed of sustainably?

### ??? Did you know ...?

It has been estimated that every year each person in the UK on average throws away seven times their own body mass in waste.

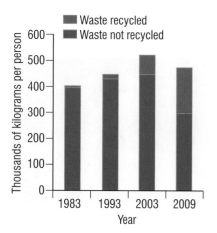

**Figure 1** Recycling of waste has increased tremendously within your lifetime

*Source: Department for Environment, Food and Rural Affairs*

**Figure 2** 15 million tonnes of waste still go to UK landfills every year

New products can increase our quality of life. Modern materials can have more useful properties than traditional materials. The industries involved in making these new products create jobs and lots of money. Nearly every new product that is made will need to be disposed of sooner or later.

### Dispose? Degrade? Recycle?

Most of the UK's rubbish ends up in **landfill sites**. Landfill is simply dumping waste into huge holes in the ground and covering it with soil. As our waste output increases, this process has become more complicated. Landfill areas are now often lined first, to prevent waste leaking into the environment. Complex systems remove or capture the liquids and gases seeping from the waste.

The UK is running out of landfill sites near the places where waste is produced. It isn't sustainable to continue using landfill sites. The materials that end up in landfill sites are just buried and lost. They are not reused. However, landfills are cheap.

**Incineration** is a way of disposing of waste by burning it. It can be more useful than landfill because it saves space and energy. After incineration, the waste is reduced to ashes, which take up a lot less space than landfills. Also, the energy released from burning the waste can be used.

The disadvantage of incinerating waste is that toxic chemicals are produced. Some of these are gases. They must be filtered out of the incinerator's emissions. The ashes that are produced are highly toxic and need to go to specialised landfills.

Some materials don't need to be processed like this because they break down naturally (**biodegrade**). They can break down because bacteria, fungi and other **decomposers** can feed on them. Most food waste is biodegradable, as are materials that come from living organisms. Synthetic materials are often not biodegradable. Biodegraded material can often be used again. For example, compost from food scraps and grass cuttings can be used as a fertiliser. However, materials only biodegrade in the right conditions. Decomposers need air, warmth and moisture to stay alive. Also, the waste products of some decomposers can be dangerous. For example, the flammable gas methane could be produced. If methane builds up inside a sealed landfill, there is a risk of explosion.

| Examples of biodegradable items | Examples of non-biodegradable items |
| --- | --- |
| Vegetable peelings | Foil wrappers |
| Paper and card | Some plastic packaging |
| Garden waste | Glass |

To avoid having to use landfill or incineration we can **recycle** or re-use as much as possible. By recycling, we can avoid having to extract the raw materials needed to make new products. However, we also create more energy demands by sorting, transporting and processing the recyclable materials. If recycling isn't performed carefully, it can use as much energy as producing materials from scratch and still cause pollution.

**a** What types of microbes help to biodegrade materials?

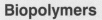

## Making better polymers

Plastic bags are made from polymers that are manufactured from products of crude oil. As well as the problems of getting rid of them, using lots of plastic bags uses up crude oil that could be used as fuel. Even though plastics make up a small percentage of the mass of waste, they take up most of the space. Reducing the volume of plastics on landfills could drastically reduce the volume of waste.

One solution to this problem is making polymers that can degrade (break down) more easily. Shopping bags and packaging can now be made from Polyvinyl Alcohol (PVOH) and Ethylene Vinyl Alcohol (EVOH). These materials are water soluble and slowly break down in the environment. However, they need to be stored carefully as humidity shortens their shelf life.

**Figure 3** A degradable plastic bag

**b** How can PVOH and EVOH help reduce the size of landfills?

## Biopolymers

Like EVOH and PVOH, biopolymers behave in a similar way to polymers from crude oil. They can be used to make shopping bags and other packaging materials. They are made from plant **biomass**, such as corn or potatoes. PLA (Polylactic Acid) is a type of biopolymer made from potato starch. It can biodegrade and doesn't use up crude oil.

Biopolymers are not perfect, however. They often have lower melting points than conventional polymers and can be less durable. Some also need specific conditions and special equipment to biodegrade them. As well as this, farmland has to be set aside to grow biopolymer crops instead of food crops.

**c** What is PLA and what is it made of?

## Photo-degradation and oxo-degradation

Additives can be used when making plastic helping them biodegrade faster:

**Oxo-degradable** plastics contain an additive that helps the plastic break down. This allows microbes to digest the plastic. Oxo-degradable plastic bags break down thousands of times faster than conventional ones.

**Photo-degradable** plastics break down when they are exposed to light. The plastic rings holding your drinks cans together have been made from photodegradable LDPE since the 1990s.

## Summary questions

1 Use the information in this spread to produce a summary table:

| Activity | Description | Advantages | Disadvantages |
|---|---|---|---|
| Landfills Incineration PVOH and EVOH | | | |

2 List the advantages and disadvantages of oxo-degradable plastics. [H]

3 Why is photo-degradable plastic a good choice for six-pack rings? [H]

### Key points

- Disposing of products in landfills and incinerators damages the environment.

- Polymers like EVOH and PVOH are used in packaging because they are water soluble and will biodegrade.

- Using biopolymers, photo-degradable and oxo-degradable materials speeds up biodegradation. [H]

231

# Summary questions

**1** Complete this table:

| Name of gas | How it is produced | Its impact on global warming |
|---|---|---|
| | When fossil fuels are burned | |
| Methane | | |
| | | Small impact – very effective at trapping heat but found in small quantities |

**2** Which one of the statements below is true? Correct the statements that are not true.

**A** Greenhouse gases have only been in the atmosphere for the last century.

**B** The Kyoto Protocol is a legal agreement for countries to reduce their output of greenhouse gases.

**C** Global warming and climate change are the same thing.

**3 a** Why do farmers use fertilisers?

**b** Rearrange the following sentences to explain how fertiliser can get into a river and what the effects might be.

- Many fish die.
- Fertiliser causes water plants such as algae to grow rapidly.
- Many plants die.
- Excess fertiliser dissolves in water in the soil.
- Fertiliser runs off fields into rivers.
- Oxygen is used up by decay-causing bacteria.
- The dead plants and algae are broken down by bacteria.
- Algae quickly cover the surface of the lake, stopping light reaching lower plants.

**c** What is this process called?

**4** Billions of plastic shopping bags are given away by supermarkets every year. Most of them end up on landfills.

**a** Describe how PVOH and EVOH can help reduce the environmental impact of shopping bags.

**b** Describe two alternatives to sending the shopping bags to landfills. Give the advantages and disadvantages of each.

**5** In addition to biodegradability we can also speed up the degradation of plastics in other ways. Explain two other methods. **[H]**

**6** Lichen can be used to detect the level of sulfur dioxide in the atmosphere.

Sulfur dioxide dissolves in rainwater to form acid rain and is released from some factories and power stations.

**a** Give the name for an organism used to detect levels of pollution.

**b** Describe how lichen can be used to show how levels of sulfur dioxide in the atmosphere change with distance from a town.

# AQA Examination-style questions

**1** Environmental scientists have been looking into the processes that release harmful gases into the atmosphere. Choose chemicals from the box to answer the questions.

> Ammonia   Carbon dioxide   Helium
> Methane   Nitrous oxide

**a** Which **two** gases are released from the combustion of fossil fuels? (2)

**b** Which gas is produced from decomposition of rubbish from landfill sites? (1)

**c** Which gas is produced due to the use of fertilisers? (1)

**d** Give the chemical formula for methane. (1)

**2** Greenhouse gases in the atmosphere are causing problems for the planet.

**a** Explain how greenhouse gases are causing global warming. (2)

**b** Suggest **one** problem caused by global warming. (2)

**3** Many things can cause pollution. Put the things that cause pollution under the correct heading in the table.

| Individual people | Industrial processes | Intensive farming | Natural |
|---|---|---|---|
|  |  |  |  |

• Artificial fertilisers
• Bacteria that break down organic matter in wetlands
• Cars
• Extraction of metals from their ores
• Herbicides
• Litter (6)

**4** Plastics made from plants are becoming more widely used. Read the newspaper article below then answer the questions.

## Biodegradable plastic – a revolutionary packaging

The applications of plant-based plastics are on the increase. One firm has developed food trays made from tapioca starch obtained from a plant called cassava. This material can withstand temperatures from −40°C to 220°C. It can also be coloured and printed. The advantage is that when this type of plastic is buried after use, or put into a composter, it breaks down naturally into water and carbon dioxide. The disadvantage is that it costs more than plastic made from crude oil. The cost of a plant-based plastic drinks bottle is up to 10% higher than regular plastic bottles.

**a** Why is it an advantage that food trays can withstand temperatures from −40°C to 220°C? (1)

**b** *In this question you will be assessed on using good English, organising information clearly and using specialist terms where appropriate.*
Discuss the advantages and disadvantages of using plants to make plastics. (6)

**c** If an oil-based plastic drinks bottle costs 79p, up to how much could a plant-based bottle cost? (4)

**5** Farmers want to grow a high yield of crops to increase profit.

**a** Which general product could a farmer put on his fields to replace nutrients so the plants will grow better? (1)

**b** Explain how this product can cause fish to die in rivers and lakes. (3)

## 15.1    Conduction

### Learning objectives

- What is conduction?
- How can we change the amount of conduction?
- Where do we make use of conduction in the home?

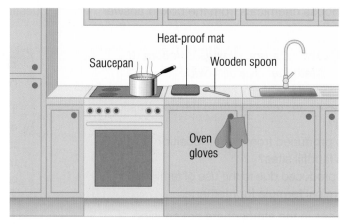

**Figure 1** Conductors of heat energy help us cook things evenly. Insulators of heat energy stop hot things causing damage.

Heat always travels from hotter places to cooler places. In solids, heat is transferred by **conduction**. The heat is transferred more quickly if:

- there is a large temperature difference
- the object has a large cross-section
- the object is made from a good heat **conductor**.

> **a** Write down **three** things that affect how quickly heat is transferred.
>
> **b** A hot cup of tea and a glass of tap water are left at room temperature. Explain which one loses or gains heat more quickly.

We cook food at high temperatures and store it at low temperatures. Fridges and freezers are lined with insulating material. This slows down heat transfers from the room into the fridge or freezer. **Insulators** are materials such as cloth, wood or plastic, which do not conduct heat well.

Oven gloves are made from padded cloth. We use them to take hot food out of the oven. The cloth, with air trapped between its fibres, is an insulator that slows down the heat transfer from the hot dish to our hand.

Many saucepans are made from metal. Metals are good heat conductors. Heat spreads quickly and evenly through a metal saucepan when it is on a hot hob. This means the food heats up quickly.

> **c** Explain which of these are good heat conductors: copper, plastic, aluminium, wood, carpet, iron.
>
> **d** Why does wrapping fish and chips in newspaper help to keep them hot?

In a solid, particles vibrate in fixed positions. Particles vibrate more in the part of a solid that is heated. The bigger vibrations pass energy onto neighbouring particles which start to vibrate more. This way, energy passes through the solid, warming up parts further away from the heat source.

Materials like metals are good heat conductors. Their particles are in fixed positions, but some electrons are free to move in the solid. When the metal is heated, the **free electrons** spread the heat quickly through the material.

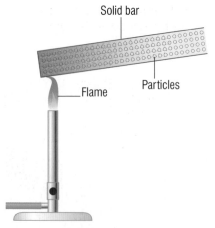

**Figure 2** Particles in a solid vibrate in fixed positions. They vibrate more when the solid heats up.

Most non-metals are insulators. They do not have free electrons so conduction is slow. Materials with trapped air pockets make good insulators. Examples of these include breeze blocks, double glazing, duvets and padded clothing. Particles in a gas do not pass energy on to each other easily.

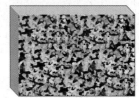

**Figure 3** Air bubbles in breeze blocks help to reduce conduction

**e** Drawing our curtains helps to keep the home warmer. Explain why.

## Activity

### Modelling conduction

To visualise why conduction happens best in a solid, line up in a row with a few of your friends and pretend you are all particles. Link elbows and, starting at one end (the hot end) one person starts moving. Soon you will find that everyone in the row cannot help but move. Next, hold hands or wrists to show particles in a liquid. When the person at the hot end moves now, their movement is not transferred as readily.

- How could you pretend to be particles in a gas? And, how will the movement be transferred this time?

Most heat is wasted in our homes through the walls, windows, floors and roofs as these have large surface areas. We can reduce these heat transfers in many ways.

- Many homes are built using breeze blocks because trapped air bubbles in the concrete reduce conduction.
- Many homes have cavity walls. There is an inner wall and an outer wall separated by a few centimetres. The trapped layer of air between the walls reduces conduction. Some people fill this with foam for extra insulation.
- Double glazed windows have two layers of glass with air trapped between them to reduce conduction.
- Carpets and rugs reduce conduction through floors.
- Loft insulation is a thick layer of padding made from fibreglass strands. It is rolled between the roof joists and reduces conduction through the ceiling.

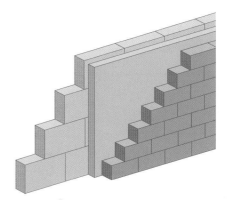

**Figure 4** Cavity walls in modern homes help to insulate them

**f** Why are heat losses from the home bigger in winter?

**g** Explain why carpets are useful for reducing heat losses in the home.

## Summary questions

1 Copy and complete using the words below:

*conductor electrons insulator nucleus vibrate*

If part of an object is heated, its nuclei will ............. more, passing energy to a neighbouring .............
Conduction is quicker in metals because some ............. are free to move.
Copper is a good heat .............
Polystyrene is a good .............

2 Explain how conduction is reduced in these cases:
   **a** A child wears several layers of clothes on a cold day.
   **b** A house with a thatched roof stays warm in winter and cool in summer.

3 Explain why metal is used to increase heat transfers in these cases:
   **a** Metal pipes carry cooling fluid through a fridge.
   **b** A baking tray is made from metal.

### ??? Did you know ...?

People can walk safely over hot coals. The skin on our feet and the ash on top of the coal conduct heat very slowly. This allows people to walk quickly over the hot coals that are at about 500 °C.

### Key points

- Conduction is when heat is transferred between neighbouring particles.

- Metals are good heat conductors.

- Non-metals and gases are poor heat conductors, or good insulators.

# 15.2 | Convection

### Learning objectives

- What is convection?
- How can we change the rate of convection?
- Where do we make use of convection in the home?

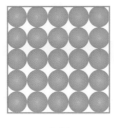
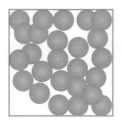
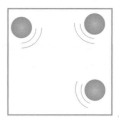

Solid          Liquid          Gas

**Figure 1** Particles are in fixed positions in solids, but can change places in liquids and gases

Liquids and gases are both **fluids**. Heat is transferred by **convection** in fluids. The particles in the fluid move, carrying energy with them.

When a fluid is warm, it is less **dense** than when it is cool. When a fluid is heated, particles gain energy and vibrate more, moving further apart and so the fluid takes up more space. The warmer fluid expands and becomes less dense. Warm water takes up more space than the same mass of cool water. This is why warm water rises and the cooler, denser water sinks. The same is true for gases such as air.

  **a** What is meant by a fluid?

  **b** How can you tell that water is more dense than cooking oil?

**Figure 2** Hot air in this balloon is less dense than the surrounding air, so it floats

**Figure 3** A wind is felt when cool air is sucked into a bonfire to replace the hot smoke

Heat is lost by convection through gaps in badly fitting window frames and doors.

Hot smoke is less dense than surrounding air so it rises up a chimney. Heat is lost up the chimney because of convection.

Fan ovens use fans to move hot air inside an oven so the food cooks evenly. Without a fan, food at the top of the oven cooks quicker because convection means the hot air rises.

The burners in a gas oven are at the base, so convection helps the heat to spread more evenly.

**c** Explain why hot air balloons rise.

**d** Where is the coolest place in a fridge?

Heat spreads quickly if a liquid or gas is heated from the base, or cooled from the top. This sets up convection currents.

The heating element is fitted at the base of a kettle. When the kettle is on, it heats up the water next to it. The warm water rises and cool water sinks to take its place. The heat circulates and spreads quickly through the water.

Some fridges have an ice box at the top of the fridge. Cooler air near the ice box sinks and warmer air rises to take its place, transferring heat away from the fridge.

**e** Where should you heat something if you want to set up a convection current?

**f** Explain how convection currents help an ice cube to cool a drink quickly.

We can reduce unwanted convection by blocking up gaps or creating small pockets of air.

People often block up old fireplaces or install draught excluders. These are strips of foam that block up gaps in window and door frames. Closing doors and windows also stops heat losses by convection.

Because of convection, warm air rises, so a lot of heat is lost through the roof of a house. Loft insulation stops the warm air reaching the loft and escaping through the roof.

**g** Write down five ways to reduce convection in the home.

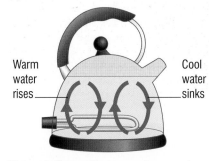

Warm water rises — Cool water sinks

**Figure 4** Convection currents spread heat evenly through the water

**?? Did you know …?**

Molten rock inside the Earth is constantly moving due to convection currents. These currents move tectonic plates, causing earthquakes and volcanoes.

**⚬⚬ links**
*For information of plate tectonics look back at 1.5 Changes in the Earth's surface.*

## Summary questions

1 Copy and complete using the words below:
*convection   fluid   rise   sink   density*

Convection is when heat is transferred through a ............

The particles move, carrying energy when ............ takes place.

The ............ of a fluid changes if its temperature changes.

Hot fluids will ............ and cool fluids will ............ if a convection current is set up.

2 Explain how you could reduce convection in these cases:
   **a** Keeping a drink hot.
   **b** Stopping convection through a letter box.

3 Explain how convection currents help a radiator heat a room.

### Key points
- Convection occurs when particles move, carrying heat with them.
- Convection takes place in liquids and gases.
- Convection currents spread heat in fluids when they are heated from the base or cooled from the top.

# 15.3 Radiation ⓚ

## Learning objectives

- What is radiation?
- How can we change the amount of radiation absorbed by a surface?
- Where do we make use of radiation in the home?

**Figure 1** Painting your house white can keep you cool in sunny weather

Many buildings in hot countries are painted white. This reduces heat transfers into the house by **radiation**, keeping the building cool.

Radiation that heats things up is also called **infrared radiation**. This radiation forms part of the electromagnetic spectrum. The energy is transferred in waves from the surface of a hot object. The waves travel at the speed of light.

Hotter objects emit (give out) more radiation in a given time from their surface than cooler objects. When radiation is emitted, objects cool down. A hot drink emits more radiation faster than a lukewarm drink, cooling down more quickly.

Radiation is absorbed by objects that are cooler than their surroundings. The objects heat up. This is why an ice cube warms up and melts if it is left at room temperature.

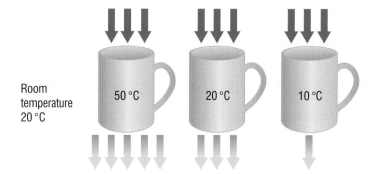

Room temperature 20 °C

50 °C   20 °C   10 °C

**Figure 2** Objects emit or absorb radiation depending on their temperature relative to their surroundings

> **a** What is another name for radiation that heats up objects?
>
> **b** Explain which of these emits more radiation? A hot drink in a white mug or a cold drink in a white mug?

Black matt surfaces are good at absorbing radiation. Solar panels on house roofs are coloured black to absorb the Sun's radiation effectively. White shiny surfaces reflect radiation and do not absorb radiation well.

All objects emit radiation if they are hotter than their surroundings. Black matt surfaces emit radiation faster than white shiny surfaces. A hot drink in a black mug emits radiation and cools down more quickly than the same drink in a white mug.

> **c** Which colour is best at absorbing and emitting radiation?
>
> **d** Explain why a refrigerated lorry painted white is more efficient to run than a dark-coloured lorry.

Some things are coloured black to increase heat transfers by radiation.

- Ovens cook food at high temperatures, so they must emit radiation easily. This is why they are black inside.
- Semiconductor chips in computers get extremely hot. They are attached to heat sinks which are painted black to increase the rate of heat loss by radiation and help keep them cool.

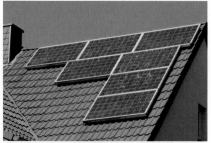

**Figure 3** Black solar panels absorb the Sun's energy efficiently

Some things are white or shiny to reduce heat transfers by radiation.

- If food is cooking too quickly, it is often covered with a layer of foil. This reflects some radiation and reduces the speed it cooks at.
- Some people fit sheets of foil behind their radiators. This reflects radiation back into the room, stopping it from passing through the wall.
- Survival blankets are shiny to reflect back the body's heat from a casualty.

**e** Why are the inside of fridges white?

**f** The insides of many saucepans are black. How does this help cook food?

**Figure 4** Black cooling fans inside equipment increase heat losses by radiation

**Figure 5** Survival blankets save lives by stopping casualties from losing heat to their surroundings

## Activity

### Modelling energy transfer

Line up with a few of your friends again. This time you need several balls (or pieces of paper) to represent energy. For conduction, one ball is passed down the line. For convection, each person picks up a ball and then moves to the back of the line to show convection current. For radiation, one person shines a torch at the person at the end of the row, missing out all the people (particles) in-between.

## Summary questions

**1** Copy and complete using the words below:

*black   infrared   radiation   surface   white*

Radiation is the transfer of energy by ............. rays.

It happens at the ............. of objects.

Hot objects emit more ............. than cool objects.

............. objects emit more radiation than ............. objects.

**2** Why does a black car heat up quicker than a white car on a hot day?

**3** Electric heaters have a curved shiny backing behind the heating element. Explain how this helps direct more heat into the room.

**4** Thermos flasks reduce heat transfers by conduction, convection and radiation. Find a diagram of a thermos flask and label it to show how these heat transfers are prevented.

### Key points

- Radiation is emitted when infrared waves transfer heat from the surface of objects.
- Hot objects radiate energy more quickly than cooler objects.
- Black matt objects radiate more energy in a given time than white shiny objects.

## 15.4

# Will you save money?

An energy performance certificate is needed if you are selling a house. This certificate shows how well energy is saved at the moment, and suggests improvements.

We have already seen some of these improvements that reduce heat losses, such as:

- draught excluders fitted round gaps in doors and windows
- hot water tank jackets to insulate the hot water tank so water stays hot for hours
- loft insulation to reduce heat losses through the roof
- double glazing to reduce heat losses through windows
- cavity wall insulation to reduce heat losses through walls. This is a layer of foam which fills the gap between a house's cavity walls.

Having effective thermostatic controls on the heating system is also important.

**Figure 1** The amount of heat lost from different houses depends on many factors

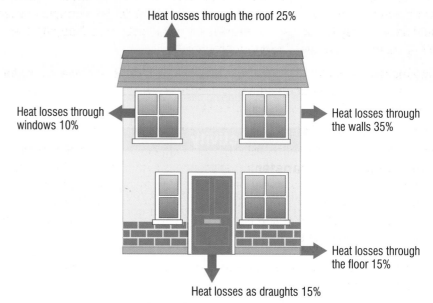

Heat losses through the roof 25%

Heat losses through windows 10%

Heat losses through the walls 35%

Heat losses through the floor 15%

Heat losses as draughts 15%

**Figure 2** This shows the amount of heat lost from different parts of a house

**a** Explain why energy performance certificates produced at different times for the same house could be different.

Identical houses in the same street have different heating bills as different energy-saving measures will be installed. The same house has different heating bills if the way the equipment is used changes. If people leave the heating on for longer, or set it at a higher temperature, their bills will be higher. Thermostats turn the heating on or off automatically to keep rooms at a constant temperature. They can be fitted to radiators so the temperature in each room can be controlled.

### Payback time

Energy-saving measures cost money to install. The time taken for the savings you make to cover the cost of installation is called the **payback time**. It is usually measured in years.

payback time = cost of installation ÷ annual savings

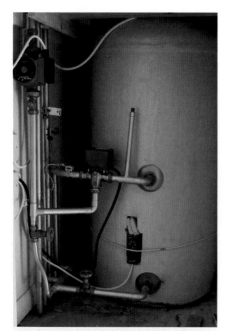

**Figure 3** Insulating a hot water tank saves money

The table shows the installation cost and annual savings for several different measures.

| Energy-saving method | Cost of installation (£) | Annual saving (£) |
|---|---|---|
| Draught proofing | 100 | 50 |
| Hot water tank lagging | 12 | 36 |
| Double glazing | 4500 | 135 |
| Loft insulation | 250 | 150 |
| Cavity wall insulation | 500 | 115 |

Using this information, the payback time for lagging a hot water tank is: 12 ÷ 36 = 0.33 years (or 4 months).

**b** Calculate the payback time for each of the measures in the table.

## Efficiency versus cost-effectiveness

Some energy-saving measures give small annual savings. If the measure is cheap and easy to carry out, it can still be cost-effective. This means that your savings are greater than the costs. A person renting a flat for a couple of years would save money if they used energy-saving light bulbs, fitted lagging over the hot water tank and used draught proofing.

Some measures would only be cost-effective if you plan to live in a house for many years. Double glazing and cavity wall insulation save a lot of money each year, but the initial cost is high. However, double glazing has other benefits: it can improve the house's appearance and cut down on noise.

Some changes reduce heating bills by improving efficiency. A boiler heats the water used in a central heating system and provides you with hot water. The efficiency of a boiler can vary between 55 and 90 per cent. Using a more efficient boiler will save heating costs by about £150 per year. However, new boilers may only last 8–10 years. If the payback time is longer than that, or the owner plans to move, this may not be cost-effective.

**c** Explain why using energy-efficient light bulbs is usually cost-effective and efficient.

**Figure 4** Some boilers are more efficient than others

### AQA Examiner's tip

There has to be a certain amount of maths in each exam paper, so payback time is one topic that you may find comes up regularly. Show all your working and do not forget the units.

### ??? Did you know ...?

Grants from local councils help householders pay to install loft insulation. This makes it more cost-effective.

## Summary questions

1 Copy and complete using the words below:

*installing loft insulation   savings   using a thermostat*

Payback time is the time taken for the ............. to match the installation costs.

............. is one way to reduce heat losses from a building.

............. is a way to use central heating more efficiently.

2 Which two energy-saving measures should be installed in these situations?
   **a** A student renting a flat for a year.
   **b** A couple living in a flat for two years.
   **c** A family living in a house for 15 years.

3 Write a paragraph explaining several ways that heating bills can be reduced without spending money.

### Key points

- Payback time = cost of installation ÷ annual savings.
- Efficient equipment is not always cost-effective to install.
- Most methods to reduce heat losses in buildings increase the insulation installed.

## 15.5 U-values

**Figure 1** The colours of a thermogram compare the heat loss from parts of a building

**Figure 2** Some building materials are better heat conductors than others

All buildings waste heat through walls, roofs, windows and floors. Different building materials lose heat faster than others. Homes lose heat more quickly if walls and roofs are not insulated.

**U-values** compare the rate of heat loss through different materials. If a material has a high U-value, heat passes through it quickly. Good conductors, such as metals, have a high U-value. Good insulators of heat, such as foam, have a low U-value.

**a** What do U-values tell us about a material?

**b** Heat passes more quickly through windows than insulated walls. Does glass have a higher or lower U-value than an insulated wall?

### U-values and building

Builders try to choose materials with a low U-value. This helps the building stay at a comfortable temperature. If it is cold outside, a low U-value reduces the rate of heat losses. More heat stays inside the building. In hot countries, heat travels into a building from outside more slowly if its materials have low U-values. The inside stays cooler.

A building using materials with low U-values needs less energy to stay at a comfortable temperature. Less energy is needed to heat the home in cold weather. As well as that, less energy is needed for air-conditioning and cooling fans in hot weather. The owner spends less money on energy bills, and there is a smaller impact on the environment. The building is more comfortable to live in too.

**c** Why are materials with low U-values good for homes in hot and cold climates?

**d** Write down two benefits of using materials with low U-values.

**Figure 3** Materials with a low U-value are good for buildings in hot and cold countries

## U-values of different materials

How quickly heat is lost from a building depends on the building materials, surface area and the temperature difference.

The table shows typical U-values of different building materials. How quickly heat is lost depends on the temperature difference and area of the material, so U-values are measured in $W/(m^2 °C)$.

| Place | Type of material | U-value in $W/(m^2 °C)$ |
|---|---|---|
| Outer wall | 22 cm solid brick | 2.2 |
| Outer wall | 28 cm brick-block cavity – insulated | 0.6 |
| Ground floor | solid concrete | 0.8 |
| Ground floor | suspended – timber | 0.7 |
| Roof | pitched with felt, 100 mm insulation | 0.3 |
| Roof | flat, 25 mm insulation | 0.9 |
| Window | metal frame, single glazed | 5.8 |
| Window | upvc frame, double glazed – 20 mm gap | 2.7 |

Walls have the largest surface areas, so the materials used for walls have a big impact on heat losses. Many buildings have two walls with a cavity (gap) between to reduce the U-value, and slow down heat losses. The building's running costs will be much lower if the walls are well insulated.

On a frosty day, the temperature difference between the outside and inside of a building can be 25 °C. This is why heat losses and heating bills are much higher in a cold winter.

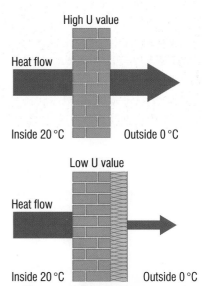

**Figure 4** Heat flows from a hotter place to a cooler place. It passes more slowly through materials with low U-values.

**e** Use information from the table. Which material should a builder use for: (i) the outer wall, (ii) the ground floor, (iii) the roof, (iv) the window frames?

**f** It is more expensive to build a home using cavity walls. Explain why this extra cost is worth paying.

### Summary questions

**1** Copy and complete using the words below:

*cooler   reduces   slower*

Heat flows from a hotter place to a ............ place.

The heat flow is ............ if the material has a lower U-value.

Insulating the wall reduces heat losses because it ............ the wall's U-value.

**2** Why is less heat lost between rooms inside a house than from walls on the outside of the home?

**3** Explain how a builder can use U-values to design an energy-efficient home. Do low U-values mean that the home is energy-efficient?

**4** This equation describes the rate of heat loss through a wall:

heat loss = U-value × area × temperature difference.

Explain why heat losses are greater through windows (U-value $5 W/(m^2 °C)$) compared with heat losses through a wall (U-value $1 W/(m^2 °C)$). Double glazed windows have a U-value of $2.9 W/(m^2 °C)$. Explain how fitting double glazing will affect the heat losses.

### ??? Did you know …?

Building regulations used to state the highest U-values allowed for building materials. Now, new homes will be carbon neutral, so materials, boilers and air-tightness must be designed to improve the building's energy efficiency.

### Key points

- U-values measure how quickly heat travels through a material.

- Low U-values mean heat is lost more slowly through the material.

- Most buildings use materials with low U-values.

# 15.6  Pollution in the home

## Learning objectives

● What substances can pollute air in the home?

● What are the symptoms of exposure to high levels of indoor pollution?

● How can air in the home be cleaned?

**Figure 1** One in every 11 children in the UK is being treated for asthma

Whether it's from car exhausts, factory chimneys or power stations, we often think of air pollution as something that happens outdoors. A lot of **air pollutants** are released into the outside environment, but many stay within homes. These substances can cause **respiratory illness**, infections and, in extreme cases, even death.

Respiratory illness is the second biggest killer in the UK (heart disease is the biggest). It affects over 8 million people. There are over 40 different types of respiratory disease. The most common include **asthma**, **hay fever** and **emphysema**.

The build up of pollutants in the home can be caused by our efforts to be more energy efficient. With air conditioners and climate control, there's often little reason to open windows. This results in air being recycled again and again – so any pollutants can't escape. Air pollution includes dust, moulds and spores (e.g. from damp walls), smoke (e.g. from cigarettes), building materials, boilers, pollen and fumes from household products.

**a**  Name two respiratory illnesses.

**b**  Why can pollutants build up in homes and offices?

Indoor air pollution affects particular parts of our homes:

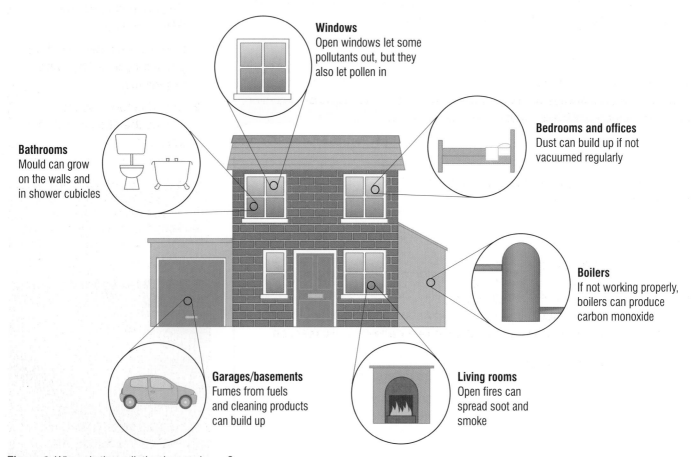

**Windows**
Open windows let some pollutants out, but they also let pollen in

**Bedrooms and offices**
Dust can build up if not vacuumed regularly

**Bathrooms**
Mould can grow on the walls and in shower cubicles

**Boilers**
If not working properly, boilers can produce carbon monoxide

**Garages/basements**
Fumes from fuels and cleaning products can build up

**Living rooms**
Open fires can spread soot and smoke

**Figure 2** Where is the pollution in your home?

## Mould and spores

**Mould** is a type of fungus that grows well in damp conditions. It reproduces by releasing millions of **spores** into the air. These spores can cause **allergies**. They can also produce poisonous chemicals called **mycotoxins**.

Mould can be killed by specialist anti-fungal bathroom cleaners. Opening windows or using an extractor fan in the bathroom is the best way to prevent mould growing in your home because it reduces humidity.

## Dust

Dust is actually made mainly of you. It's formed from the dead flakes of skin cells you shed all the time, as well as fibres from clothing and furnishings. Dust can trigger allergies, itching and asthma attacks. It can also provide food for **dust mites**. Every home has dust mites living in it but if there are too many of them their waste can also cause allergic reactions. The static electricity on a TV screen can attract dust easily, so anti-static wipes and cleaners are a good way to keep dust at bay.

**c** How can dust mites be harmful?

## Soot and smoke

Soot and smoke are both made up of microscopic carbon particles that can permanently damage the **alveoli** in your lungs. This condition is called emphysema. The main symptom is shortness of breath. Very few homes in the UK now use wood fires. Gas-burning fires hardly produce any soot. However, passive smoking remains an issue in some homes.

## Fuels and cleaning products

These can cause chemical damage to the respiratory system and even poison you. They should always be used in well-ventilated areas. Signs of respiratory damage include tiredness, nausea and headaches.

## Carbon monoxide

**Carbon monoxide** is a gas that can be given off by faulty boilers. Carbon monoxide poisoning can kill.

## Pollen

**Pollen** is a reproductive cell released by plants. Like mould spores, it can cause allergic reactions. Pollen allergies are called hay fever and affect around 20 per cent of people in the UK. Different people are allergic to different types of pollen. Keeping windows closed on hot days can stop pollen getting into your home. **Antihistamines** can be taken to reduce the allergic reaction.

**d** What causes hay fever?

**Figure 3** Dust mites live off your dead skin cells

### ∞ links

*For more information about carbon monoxide poisoning see 15.8 The silent killer.*

## Summary questions

1 Summarise the information on this page as a table with the headings; 'Area in home', 'Type of pollution', and 'Harm caused'.

2 Name three types of indoor pollution caused by living organisms.

3 Describe four ways to reduce the effects of indoor pollution.

## Key points

- Closed air systems in homes can lead to a build up of indoor pollutants.

- Indoor air pollutants include dust, mould and spores, smoke, fumes from household products, carbon monoxide and pollen.

- Some of the symptoms of exposure to high levels of indoor pollution are asthma, headaches, tiredness, dizziness, nausea, an itchy nose and/or a sore throat.

## 15.7    Household hazards

### Learning objectives

- What household products can be dangerous?
- How are hazards labelled in the home?

**Figure 1** Hazards in the home

We use a huge range of chemicals in the home in order to make our lives easier. Often, these chemicals can be harmful to us. By law, hazard symbols must be added to the labels of products if they contain a dangerous ingredient. This is so consumers are aware of the risks and can take the right precautions.

**a** Why are hazard symbols added to product labels?

Each different type of hazard has a symbol so it is easily recognised (see Figure 2).

**b** What is the difference between harmful and irritant substances?

**c** Why is nail polish remover hazardous?

### What happens if I'm exposed to a hazardous chemical?

Different chemicals affect the body in different ways. This can make it difficult to tell what someone has been exposed to. Symptoms could include:

- dizziness
- nausea (feeling sick)
- headaches
- rashes
- itching
- tiredness
- sore nose/throat.

**d** What might happen to your skin if you spilled bleach on it?

### How can I make sure I am safe?

Some simple precautions can reduce the risk of you being harmed by chemicals in the home:

- Always read the label – hazard symbols will let you know what the danger is.
- Avoid exposing your skin to any chemical you are unsure of.
- Use potentially dangerous chemicals in well-ventilated areas. Don't let fumes build up.
- Use a disposable face mask when working with chemicals that can damage your airways.
- Never transfer a chemical to another container – someone else might not know what it is.
- Use eye protection and cover your clothing if you might get splashed by a hazardous chemical.

### ?? Did you know ...?

Sometimes, hazardous materials end up in household products by mistake. In 2008, thousands of people were harmed by gases being released by sofas. The gases were produced by a chemical used to prevent mould. In 2010, a total of 20 million pounds was paid in compensation to victims of the sofas. This was the highest ever compensation payment of its kind.

| Symbol | Risks and precautions | Examples |
|---|---|---|
| Harmful | Harmful substances can damage your health. If inhaled, they might make it difficult to breathe. If accidentally swallowed, they could cause vomiting, stomach pains or diarrhoea. Harmful substances can also damage other organ systems, like your nervous system. Wash immediately after skin contact and seek medical attention if swallowed or inhaled. | Some laundry additives are harmful if swallowed. |
| Irritant | Irritants can cause damage to skin, eyes and mucous membranes (i.e. breathing passages). This damage may be in the form of a rash, itching or blisters. Irritation of the respiratory system could cause coughing, make it difficult to breathe, or cause asthma attacks. If you are working with an irritant, use a mask and gloves. | Bleaches irritate the skin. They can also produce toxic chlorine gas if mixed with acids. |
| Corrosive | As the picture suggests, corrosive substances can 'eat through' other materials, including human flesh. Not all corrosive substances are liquids. Some solids can also burn through your skin. You should never touch a substance with this label on it. Use gloves when working with corrosive chemicals. Medical advice should always be sought after a chemical burn. | Oven cleaners are often corrosive. |
| Environmental hazard | Environmental hazards pose a danger to the living organisms in the environment if they are not disposed of properly. They are usually harmful to humans as well. Some materials have to be disposed of in a particular place (i.e. a waste oil bin in a recycling centre) because it is illegal to put them in with domestic waste. | Disinfectants and drain cleaners can be dangerous to the environment if they are spilled near living organisms. Drain cleaners should be used carefully outdoors in case they pollute places where things are permitted to grow. |
| Flammable | Flammable substances are easy to ignite. They don't always need a flame or a spark. Sometimes enough heat can make a flammable substance ignite on its own. | Nail polish removers and other solvents are often flammable. |
| Toxic | Toxic substances can poison you by seriously damaging one or more organ system in your body. Toxic substances aren't just poisonous to eat or drink. Some toxic substances can be absorbed through your skin, lungs or even eyes. Always read the instructions very carefully and get medical advice immediately if you suspect someone has been poisoned. | Very few products in the home are toxic. However, many products in the home will release toxic gases if they burn. In house fires, far more people are killed by toxic smoke than by the actual fire. |

**Figure 2** Hazard symbols, their meanings and examples from home

## Summary questions

1 Read the 'Did you know?' box and answer the following questions:
   a Why had the harmful compound been added to sofas?
   b Name a symptom that the victims experienced?

2 Fly spray can kill fish in aquariums if they aren't covered up. What hazard symbol would you expect to see on fly spray?

3 What safety precautions should you take if cleaning something with bleach?

## Key points

- Many household products contain hazardous chemicals.
- Products must be clearly labelled with hazard symbols.
- Exposure to hazardous chemicals can cause dizziness, rashes, headaches and other symptoms.

## 15.8

# The silent killer

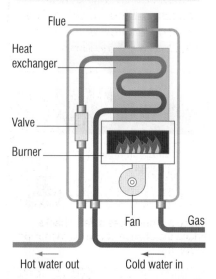

Figure 1 A modern gas boiler

∞ **links**

*For information about complete and incomplete combustion look back at 8.2 Burning fuels.*

Most of the energy we use in homes is supplied by electricity, but some comes from fuels. Methane gas is burned in domestic boilers to heat water (see Figure 1).

## What is incomplete combustion?

In a modern boiler the methane gas goes into the boiler and a fan mixes it with air, which contains oxygen. The methane combusts and the energy released warms up the water. Waste gases escape through a pipe at the top (called a **flue**). Normally, the methane is completely **combusted**. The reaction happening is:

methane + oxygen ⟶ carbon dioxide + water

However, sometimes things can go wrong. If the flue becomes blocked, waste gases will not be able to leave. Also, if the fan breaks, not enough oxygen gets in.

If there is not enough oxygen, the methane isn't able to burn properly. This is called **incomplete combustion**. When this happens, **carbon monoxide** (CO) and carbon (soot) are produced as well as carbon dioxide. The reaction is:

methane + (not enough) oxygen ⟶ carbon dioxide + carbon monoxide + carbon + water

Incomplete combustion doesn't release as much energy as complete combustion. So the boiler is less efficient. It is also very dangerous because carbon monoxide is a toxic gas.

### Practical

**Bunsen burners and incomplete combustion**

You can use different Bunsen burner flames to heat the same volume of water for a minute. Record the temperature change of the water. Draw a graph of temperature change against the size of the air hole. You can use this graph to describe the relationship between the energy released from burning methane and the amount of oxygen available for combustion. Think carefully about how to make this a fair test.

**a** What are the products of incomplete combustion?

## Why is carbon monoxide so dangerous?

Every year around 50 people in the UK are killed in their homes by carbon monoxide poisoning. It is a colourless, odourless and tasteless gas, which makes it very difficult to detect. Carbon monoxide is dangerous because it replaces oxygen in your blood.

Normally, your **red blood cells** carry oxygen around your body. They collect it in your lungs and drop it off in the tissues that need it. They contain a protein called **haemoglobin**, which holds onto oxygen. Unfortunately, haemoglobin holds onto carbon monoxide over 200 times as tightly as oxygen. If you breathe in a lot of carbon monoxide, it will take the place of oxygen in your red blood cells. Slowly, you will become starved of oxygen and may even die.

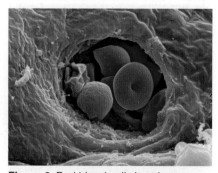

Figure 2 Red blood cells bond more strongly to carbon monoxide than to oxygen

**b** Why is carbon monoxide difficult to detect?

## How do I know if I've been poisoned by carbon monoxide?

The symptoms of poisoning depend on the concentration of carbon monoxide and how long you are exposed to it. At very low concentrations, carbon monoxide poisoning causes mild headaches and dizziness. At high concentrations it can lead to unconsciousness and death in minutes. The longer you are exposed to it, the worse the symptoms get.

These are some of the symptoms of carbon monoxide poisoning from a domestic boiler:

Because the symptoms of CO poisoning are quite mild at low concentrations, it often gets mistaken for a cold.

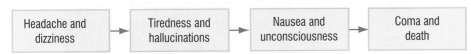

**Figure 3** Symptoms of carbon monoxide poisoning

## Preventing carbon monoxide poisoning

Servicing boilers every year can help prevent them producing carbon monoxide. A **gas service engineer** will check that everything is working properly. The engineer will perform a flue test to detect any dangerous emissions. A faulty boiler may also have black soot marks around it. This is because incomplete combustion also produces carbon.

Carbon monoxide detectors are available to use in your home. Some work in a similar way to smoke detectors and some contain a chemical that changes colour in carbon monoxide.

**c** What could black soot marks around a boiler mean?

If you think your boiler is producing carbon monoxide it is important to get to a source of fresh air immediately. Then arrange for a gas service engineer to check it. Mild carbon monoxide poisoning can be treated by breathing normal air for a few hours. More serious cases are treated by putting the victim into a **hyperbaric chamber**. This is pressurised with oxygen, which helps force the carbon monoxide out of the victim's red blood cells.

### Did you know … ?

Carbon monoxide poisoning can cause hallucinations. This has been responsible for several people thinking their house was haunted. They were, in fact, being poisoned.

### AQA Examiner's tip

You will remember carbon monoxide from Unit 2 when you learnt about smoking. The situation is a new one but the effects are still the same.

**Figure 3** Annual gas safety checks reduce the chance of carbon monoxide poisoning

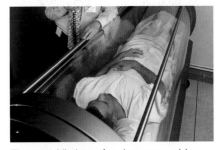

**Figure 4** Victims of carbon monoxide poisoning are treated in hyperbaric chambers

## Summary questions

1 Copy and complete using the words below:

*oxygen  colourless  water  monoxide  carbon  burn  incomplete*

Domestic boilers combust (............) methane to release energy. Complete combustion produces carbon dioxide and ............, but incomplete combustion produces carbon ............ and ............ as well as carbon dioxide. If there is not enough oxygen, ............ combustion occurs. Carbon monoxide is a ............, odourless poison. It prevents blood from carrying ............, so can be deadly.

2 Which two factors affect the seriousness of carbon monoxide poisoning?

3 Describe two ways to prevent carbon monoxide poisoning and two ways to treat it.

### Key points

● Incomplete combustion happens when a fuel is burned without enough oxygen.

● Incomplete combustion of a fuel can produce carbon monoxide and carbon.

● Carbon monoxide is poisonous because it replaces the oxygen in your red blood cells.

# 15.9 Radon gas

**Figure 1** Granite rocks naturally produce radon gas

Lung cancer is a killer. Smoking is the most damaging thing you can do to your health, increasing your risk of lung cancer enormously. Second-hand smoke can also cause cancers in non-smokers. The next most common cause of lung cancer is radon gas.

**Radon** gas is a radioactive gas, produced naturally from uranium in rocks. It is colourless and odourless, and seeps from the ground and into our homes. It is not usually a big health risk to most people. Since cells in a smoker's lungs are already damaged, the radon can be more harmful, making lung cancer more likely.

**a** Where does radon gas come from?

**b** Why is radon gas a health hazard?

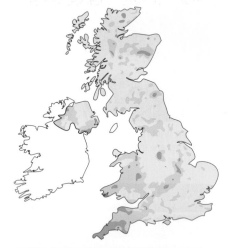

**Figure 2** Radon occurs in different amounts across the UK. The darker the colour on the map, the higher the level of radon.

*Based on information supplied by The Department for Environment, Food and Rural Affairs.*

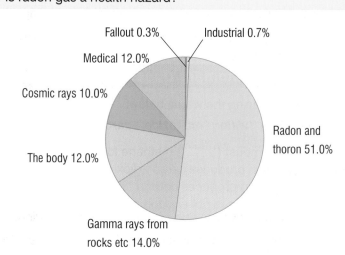

Fallout 0.3%   Industrial 0.7%
Medical 12.0%
Cosmic rays 10.0%
Radon and thoron 51.0%
The body 12.0%
Gamma rays from rocks etc 14.0%

**Figure 3** The main sources of radioactivity we are exposed to every day

We are naturally exposed to radioactivity all the time (see Figure 3). This includes radioactivity from building materials, soil, cosmic rays and food. The background radioactivity is usually too low to be a health risk. The amount of radon gas in your home depends on where you live. Some parts of the country have higher levels than others. If the rocks and soil below your home contain large amounts of uranium or plutonium, it is likely more radon gas will seep inside.

This is a problem if the gas is allowed to build up inside the home. Over many years, the radon can damage cells in a person's lungs. Some people will develop lung cancer as a result. If the house is well ventilated, the radon gas cannot build up. This means that the health risk is very low.

**c** Where does radioactivity come from naturally?

**d** How can you prevent radon gas becoming a problem in the home?

## Reducing radon levels

The amount of radon in a building can be tested using small detectors. Most homes in the UK have a radon level of 20 Bq per cubic metre. If the levels are above 200 Bq per cubic metre, the health risks are greater and action should be taken.

It is possible to reduce the level of radon in buildings. However, this can be too expensive unless radon levels are high. The risks of radon are very small in the short term, but increase over long periods of time. Reducing radon levels is important if people are smokers or exposed over long times.

In some areas of the country, it can be hard to sell a home that may have high radon levels unless action has been taken to reduce the levels. Businesses may need to control their radon levels to protect the health of their workers.

The three main ways to reduce radon levels are to:

- seal the floor so radon cannot seep in from the ground underneath
- increase the ventilation so the radon cannot build up inside
- use an extractor fan under the building to pump the radon away before it enters the building.

The best method depends on the level of radon and type of building.

**e** Write down three ways to reduce the radon level in a building.

**f** Explain three reasons why some people do not take action to keep radon levels below a certain level.

**Figure 4** Radon tests help people decide whether to reduce radon levels in their home

## ∞ links

*For information on radioactivity look back at 11.7 What is radioactivity?*

---

### Summary questions

1 Complete the following sentences using these words:

*cancer gas uranium ventilation*

Radon is a radioactive ............ that can cause ............ .

Radon seeps from rocks that contain ............ .

You can reduce radon levels by increasing ............ in a building.

2 Explain two reasons why radon is not a problem for most people.

3 You have received a letter saying that the radon levels in your home are above 200 Bq per cubic metre. Explain what you can do to reduce the levels of radon in your home.

### Key points

- Radon is a radioactive gas that can cause cancer.
- Radon seeps from rocks under buildings that contain uranium and plutonium.
- Radon levels can be reduced by ventilating the building.

# Summary questions

**1 a** Explain how heat is transferred in solids by conduction.

**b** Write down **three** places where conduction is reduced in a building. What materials are used as insulators?

**2** Complete these sentences using the words below. You can use the words more than once:

*conduction   convection   radiation*

The inside of a fridge is white to reflect ............ and reduce heat transfers.

A double glazed window has a layer of air trapped between two panes of glass. This reduces ............

Warm air from a radiator rises and circulates round a room. This is ............

Draught proofing reduces heat losses by ............

**3** Complete the gaps in this table using some or all of these words:

*black   white   shiny   dull*

| | Best colour | Best type of surface |
|---|---|---|
| Absorbing radiation | | |
| Emitting radiation | | |
| Reflecting radiation | | |

**4 a** Explain **two** reasons why many people will save money by installing loft insulation. It costs £250 to insulate a typical home and the savings are likely to be £150 per year.

**b** Give **two** reasons why some people would not install loft insulation.

**c** Explain why a builder should always aim to use materials with the lowest U-value, even if these may be more expensive initially.

**5** Name some pollutants you might find in the following parts of a home, and describe how they can harm you.

**a** Near windows

**b** Offices and bedrooms

**c** Bathrooms

**6** The labelled diagram shows many pollutants in the home.

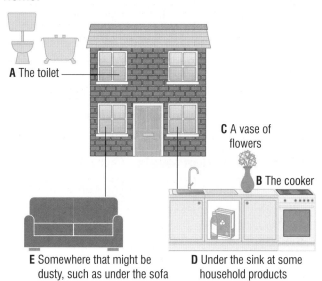

**A** The toilet

**C** A vase of flowers

**B** The cooker

**E** Somewhere that might be dusty, such as under the sofa

**D** Under the sink at some household products

**a** For each pollutant in the list, suggest a letter from the diagram that might be responsible.

**i** Pollen

**ii** Smoke

**iii** Dust

**b** Name **two** common symptoms of exposure to high indoor pollution levels.

**c** Suggest and explain a simple method of reducing the level of indoor pollution.

**7** Suggest which hazard symbol should appear on containers of the following substances.

**a** Petrol

**b** Bleach

**c** Oven cleaner

**d** Rat killer

**8** Carbon monoxide poisoning kills more than 50 people in the UK every year.

**a** What household item can produce carbon dioxide if it isn't working properly?

**b** What is the name of a common chemical reaction that can produce carbon monoxide?

**c** Why is carbon monoxide poisonous?

**9** Radon is a radioactive gas that seeps from rocks. It builds up in homes and can cause an increased risk of lung cancer for people living in some houses.

**a** Explain why people may not be aware of the risks of radon gas.

**b** Explain why people may be unwilling to install measures that prevent the build up of radon gas.

# AQA Examination-style questions

**1** Match the type of heat transfer to its definition.

| Heat transfer | Definition |
|---|---|
| Conduction | An electromagnetic wave |
| Convection | Heat energy is passed to neighbouring particles |
| Radiation | Occurs because of the movement of particles in liquids and gases |

(2)

**2** The table shows three different ways to insulate walls.

| Method of insulation | Cost in £ | Annual saving in £ | Payback time in years | $CO_2$ saved per year in kg |
|---|---|---|---|---|
| Solid wall | 10 500 | | 26.25 | 2100 |
| Cavity wall | 250 | 115 | | 610 |
| Thermal wall paint | | 25 | 4 | 75 |

**a** Complete the table. (3)

**b** *In this question you will be assessed on using good English, organising information clearly and using specialist terms where appropriate.*
Write a conclusion based on the data in the table, suggesting with reasons which method of insulation is the best. (6)

**3** The 'European Agreement concerning the International Carriage of Dangerous Goods by Road' is in charge of making sure hazardous substances are transported safely. They have several classes of materials.

- Class 1 Explosive substances and articles
- Class 2 Gases
- Class 3 Flammable liquids
- Class 4.2 Substances liable to spontaneous combustion
- Class 5.1 Oxidising substances
- Class 6.1 Toxic substances
- Class 7 Radioactive material
- Class 8 Corrosive substances

**a** What class would be given to a crate of rat poison carrying this symbol? (1)

**b** What class would be given to a tanker of liquid pentane carrying this symbol? (1)

**c** What class would be given to a lorry containing waste material from a nuclear power station? (1)

**d** What class would be given to a lorry carrying bottles of hydrochloric acid with this symbol? (1)

**4** Choose the correct word from each box to finish the sentences.

**a** When fitting a domestic boiler, care has to be taken to ensure a good supply of ............. .

*air   electricity   water*

(1)

**b** Otherwise ............ combustion occurs.

*full   incomplete   radioactive*

(1)

**c** This leads to the formation of ............ products including ............ .

*infectious   penetrating   toxic
carbon monoxide   methane
nitrous oxide*

(2)

**5** Use the equation to answer the questions.

$$\text{efficiency} = \frac{\text{useful power out}}{\text{total power in}}$$

**a** How efficient is a new wall-mounted gas boiler that uses 28.00 kW and converts 25.62 kW into useful energy for heating? (2)

**b** How efficient is a similar boiler that sits on the floor and transfers 30.60 kW and wastes 3.65 kW of energy? (4)

**c** Choose a reason why the second boiler is not as efficient as the first.

| Energy transfer through the floor | Energy transfer through the walls | Needs more power to run |
|---|---|---|

(1)

**1** Drinking alcohol increases the risk of liver sclerosis.

The graph shows the number of women that died from sclerosis of the liver in different parts of Europe from 1955 to 2000.

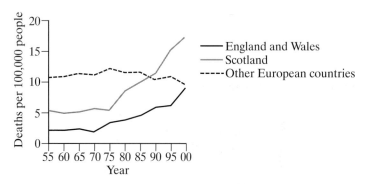

**a** Which part of Europe had the biggest change over the time period and by how much did it change? *(2)*

**b** In what year were the number of deaths from sclerosis of the liver the same in Europe and Scotland? *(1)*

**c i** Leeds in the north of England had a female population of 369 570 in 2000. Estimate how many of them would have died that year from sclerosis of the liver. *(2)*

**ii** How do you expect this number to change in the future? *(1)*

**d** Suggest **three** things that you can say about the drinking habits of other European women when compared with Scottish women from 1955 to 2000. *(3)*

**2** *In this question you will be assessed on using good English, organising information clearly and using specialist terms where appropriate.*

Our white blood cells defend us against microbes entering our bodies.

**a** Describe the stages that follow once a lymphocyte encounters a pathogen. *(6)*

**b** Explain why a small number of bacteria that get through a cut in your skin into your body can make you feel ill. *(2)*

**3** Some drugs can be used to improve the quality of our lives.

**a** Match the medical drug to its use.

| Medicinal drug | Use |
|---|---|
| Aspirin | Kills bacteria |
| Paracetamol | Anti-inflammatory |
| Antibiotic | Painkiller |

*(2)*

**b** Give **two** reasons why doctors are worried about the over-prescription of antibiotics. *(2)*

**4** Many household products have hazard symbols on their packaging.

**a** Match the hazard symbol to its meaning.

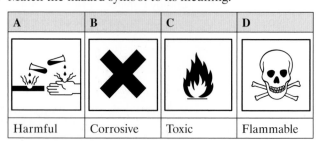

*(4)*

**b** Another danger in the home is radon gas.

**i** Why might there be radon gas in a home? *(1)*

**ii** Suggest **one** reason why radon gas can be dangerous. *(1)*

**5** Materials scientists are constantly looking for new products to make our lives easier.

  **a** Thermochromic materials change colour according to their temperature.

    Suggest a reason for using thermochromic materials in each of these cases.

    **i** A baby's bottle *(1)*

    **ii** A t-shirt *(1)*

    **iii** A bottle of milk *(1)*

    **iv** A ceramic hob *(1)*

  **b** Give the name for a material that changes colour according to how much light there is. *(1)*

**6** Some scientists want to use cloning techniques to make new organs for humans. The new organs will cut down waiting times and hopefully reduce the numbers of organs that are rejected by their new host.

  Cloning is done by removing the nucleus from a donor egg, and replacing it with the DNA from the organism to be cloned.

  **a** What is a clone? *(1)*

  **b** Suggest **two** reasons why some people might be against cloning organs in this way? *(2)*

**7** Lichen is an indicator species that is very sensitive to sulfur dioxide, which causes acid rain.

  There are many different varieties of lichen. There are those that are very hardy nearest the town centre. They are green and crusty. Further from the town more types of lichens can grow that look leafy and shrubby.

  The table shows the number of different varieties found growing on walls at different distances from the town centre.

| Distance from town centre in km | Number of different varieties of lichen found |
|---|---|
| 0 | 0 |
| 2 | 15 |
| 4 | 13 |
| 6 | 15 |
| 8 | 25 |
| 10 | 27 |
| 12 | 46 |
| 14 | 46 |
| 16 | 65 |

  **a** Explain at which distance from the town there is an odd result. *(2)*

  **b** Write and explain a conclusion based on the results. *(2)*

# Planning and risk assessing

Your Controlled Assessment is worth 25 per cent of the total marks for your Science GCSE. It consists of **one** practical investigation based on themes from the specification.

Each task assesses 'How Science Works' skills and you will need to research an application of the science context. You are free to use methods and techniques that are not mentioned in the specification.

You will need to produce a record of your investigation that shows your skills in:

- research
- planning
- assessing and managing risk
- collecting data
- processing data
- analysing data
- evaluating the practical activity.

The following pages on the Controlled Assessment unit are designed to help you develop your 'How Science Works' skills and enable you to achieve the best mark possible.

## Why is planning important?

Imagine going about your daily life without making any plans. You arrive at school but have not planned your day properly. You arrive with no books, no pencil case and no lunch. Your day is unlikely to go well. It is also important to plan a scientific investigation properly. Then everything will go as well as it can and all steps of the investigation will be done in the correct order. If you try to make a cake but miss out ingredients or add them in the wrong order, your cake will not turn out right. It is better to follow a recipe step by step.

You will be asked to write a plan for your investigation, which includes saying what you are doing and why. The more detail you give and the more accurate your plan is, the better your mark will be. You must write your plan in the **future tense** as if you are *going* to do your experiment, not as if you have *done* it. Alternatively you can write it in bullet points like following a recipe in a cookery book.

**Figure 1** Watch out for hazard symbols on chemicals you use and think about the risk of using these chemicals

## ⚭ links

*Look back to 15.7 Household hazards for more about hazard symbols.*

## Safety first!

In industry, employees are required to follow Standard Operating Procedures or SOPs for all experiments or investigations. SOPs are important as they help get reproducible results across different groups/organisations carrying out the same practical work.

Scientists need to risk assess the experiment to identify all of the risks and make sure they know how to control them. To do this, a risk assessment is performed. This involves looking at all of the equipment, chemicals and techniques and identifying the risks (what could go wrong) and the control measures (how to prevent accidents). There are different ways to write a risk assessment but the easiest is to use a table with the following headings:

| Equipment, chemical or technique (What is the possible hazard? What could happen?) | Risk (What are the chances of harm happening? How serious could it be?) | Control measures (How can you prevent something happening?) | Emergency action (What first aid action would you take if something did happen?) |
|---|---|---|---|
|  |  |  |  |

## Activity

### Practising planning and risk assessing

It is important to practise the skills needed before you complete your centre-assessed unit. Think about something you do on a regular basis, for example, making a cup of tea.

- Write a step-by-step plan for making a cup of tea. Put all steps in the correct order and either in **future tense** or in bullet form.
- Now write a risk assessment for making a cup of tea using the table headings above. Remember to be as detailed as possible.

## Activity

### Preparing calcium carbonate – an antacid

#### Procedure:

- Weigh out 11.0 g of calcium chloride crystals and place in a 250 cm$^3$ beaker. Add 50 cm$^3$ of distilled water and stir to dissolve.
- Weigh out 15 g of anhydrous sodium carbonate and place in another 250 cm$^3$ beaker. Add 100 cm$^3$ of distilled water.
- Pour the sodium carbonate solution into the calcium chloride solution, stirring all the time.
- Filter the mixture by pouring through a filter funnel into a 250 cm$^3$ conical flask.
- When all the mixture has been filtered, wash the calcium carbonate in the funnel with a small amount of distilled water.
- When all of the liquid has run through, take the filter paper out of the funnel and lay it flat on a paper towel. Put it in an oven to dry.

#### Tasks

1 Highlight all of the chemicals that are hazardous. Remember you will use some and produce one too!
2 Highlight all of the apparatus that could be hazardous.
3 Read through the method one more time. Are there any other pieces of equipment or requirements that have not been mentioned but should be included on your risk assessment?
4 Now produce a risk assessment for the procedure.

## Summary questions

1 Why is it a good idea to plan a practical before you start? Give two reasons.
2 Explain why another person should be able to repeat your experiment.
3 Explain the difference between a hazard and a risk.

## AQA Examiner's tip

### What to include in your plan

- The aim of the investigation
- The variables you will study in the investigation
- How you will make the investigation fair and safe
- Equipment to be used
- Measurements/readings to be taken (how many, over what range?)
- Table to record your results

## Key points

- Planning is important before you start your controlled assessment so you know what you are doing and what equipment you will need.

- As part of your plan you should assess all risks so you can make the plan safe.

# Collecting and recording evidence

## Learning objectives

- How can I record and present my data from an investigation in a suitable table?

- Why should I repeat my measurements or observations?

- What is the most appropriate type of graph to use to present my data?

## Planning a table of results

Imagine trying to do your homework with all of your books, pens and pencils scattered all over your desk. You are unlikely to find everything you need and you may even lose your homework amongst all of the chaos. When you carry out an experiment, it is a good idea to plan your results table before you start so that, when you record your results, they are well ordered and you can find everything.

The first step in planning a results table is to decide what you are going to measure in the investigation. For example, if you are boiling some water in a beaker and you want to measure the temperature, you would need to record the temperature over a certain time interval. You then need to decide how many times you are going to do the experiment and include enough columns in your table. If you have repeated your experiment, it is a good idea to include enough columns for each test result and a column for means (averages). The final step is to make sure that you have included headings and units.

The more organised your table is, the easier it is for you to record your results, analyse your findings and for someone else to follow what you have done as well.

### Activity

**Practise planning a table**

Imagine that you are going to carry out an experiment to find the best material to use to make a saucepan. You are going to boil 100 cm³ of water in four different materials until the water reaches 100 °C. You are going to do this for glass, copper, steel and aluminium.

Plan a table to record your results. Think about:

- what you need to record
- the headings and units you need to include
- how many times you will do the experiment
- whether you need to include a column for the means.

AQA Examiner's tip

Remember that:

A categoric variable is described by words. It has certain fixed values. Examples are 'the type of metal' or 'the location of the water sample'.

A continuous variable can have any numerical value. So anything measured is a continuous variable. Examples are 'time', 'temperature' or 'distance'.

## Why repeat measurements?

It is always good practice to repeat measurements to check your findings and to see if your results are repeatable and precise. The more times you can repeat an experiment, the more accurate your findings are likely to be. By repeating, you can check whether the results agree or are close enough to show that they are precise. For example, if you time how long it takes to boil 50 cm³ of water and find that the first time it takes 38 seconds, the second time it takes 42 seconds and the third time it takes 1 minute and 14 seconds you should be able to spot an **anomalous result**. There is always a little bit of variation in any repeat measurement. However, this is a result that does not fit the pattern and usually happens because a mistake has occurred. Which is the anomalous result in the set of three repeat readings above?

It is always a good idea to repeat measurements if you find anomalous results so that all of your results in a repeat set are close together (precise). If you haven't got time to repeat an anomalous reading it can be discarded – identify it, then don't take it into account when calculating the mean (average).

## Presenting data in a graph

Depending on the data you have recorded, you can present your data in different forms. A graph is an easy way of analysing your findings very quickly.

A **bar chart** is used for categoric data; for example, showing the density of different metals.

A **line graph** is used for continuous data; for example, showing how temperature changes over time. If you can, you should draw a line of best fit which is a line that goes through the middle of the distribution of all the points. (Don't forget that a line of best fit can be a curve or a straight line.)

The **independent variable** is the variable that you choose to vary during the investigation. This is plotted on the $x$ (horizontal)-axis. The **dependent variable** is the one that you measure to judge the effect of varying the independent variable. This is plotted on the $y$ (vertical)-axis.

Choosing an appropriate scale for your graph is very important. The scale should allow you to clearly see your graph. You should try to use as much of the graph paper as possible but choose a scale that is easy to read. For example, do not choose a scale with 10 squares for each 30 units. (Imagine trying to plot a value such as 17.4 units on that scale!) You should also make sure that your scale is linear. This means that each square on the graph represents the same quantity. For example, if you are plotting a graph of temperature against time, each square for time should represent the same number of minutes or seconds.

## Summary questions

1 Why is it important to include headings and units in a results table?

2 Why might you repeat certain measurements within your experiment?

3 What type of graph would be used to display the results of the following investigations?
   a How does temperature affect the speed of a tennis ball?
   b Which type of supermarket carrier bag is the strongest?

4 Look at the table above. The student tested a fifth sample of water from a canal. The results of the three tests were 18.4, 18.8 and $18.6\,cm^3$. What is the mean of these three readings?

## Key points

- When planning an investigation you should also produce a table to record your data in.

- Repeat any measurements or observations a number of times to improve the accuracy of your results.

- A bar chart is used for categoric data.

- A line graph is used for continuous data.

## Activity

### Spot the odd one out

Some students have carried out an experiment to look at acidity in water taken from four different reservoirs. They have recorded how much water from each reservoir was needed to neutralise $25\,cm^3$ of sodium hydroxide. Their results table is shown below.

| Reservoir | Volume of water needed to neutralise $25\,cm^3$ of sodium hydroxide ($cm^3$) | | | |
|---|---|---|---|---|
| | Test 1 | Test 2 | Test 3 | Mean |
| Matlock Dye Works | 28.1 | 28.6 | 29.3 | 36.3 |
| Butterley Reservoir | 62.5 | 44.2 | 60.2 | 38.8 |
| Carsington Water | 20.2 | 21.7 | 22.4 | 30.5 |
| Ratcliffe on Soar Power Station | 14.3 | 52.5 | 51.6 | 39.5 |

1 Which results are anomalous? How do you know?
2 What would be the best way to present these findings in graph form?
3 Draw a graph of the results. Remember to use appropriate scales, label your axes and include units.

# Analysing and evaluating evidence (k)

## Learning objectives

- How can I draw conclusions from evidence collected in my own investigation or from secondary sources?
- How can I evaluate my investigation?

## Looking for patterns in data

After carrying out your practical work, you will have a table of your results and a graph to display the data collected. Your task will then be to analyse what the data can tell us about the question you gathered evidence to answer.

Line graphs can tell us the relationship between the two key variables investigated. How are they linked, if at all? Here are some patterns you need to look out for:

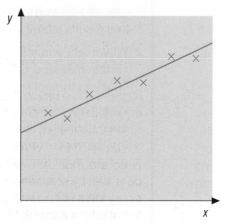

**Figure 1** This line of best fit shows a positive linear relationship. We can say, 'As $x$ increases, $y$ increases at a constant rate.'

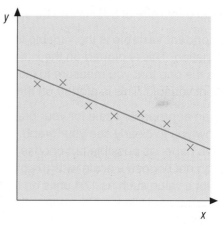

**Figure 2** This line of best fit shows a negative linear relationship. We can say, 'As $x$ increases, $y$ decreases at a constant rate.'

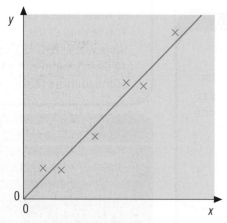

**Figure 3** This line of best fit shows a special positive linear relationship where the line starts at the origin (0, 0). This relationship between $x$ and $y$ is called directly proportional. In this case we can say, 'If we double $x$, then we also double $y$.'

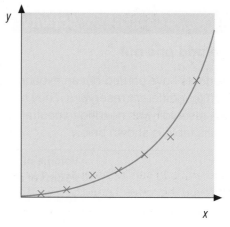

**Figure 4** This line of best fit shows a positive relationship so we can say, 'As $x$ increases, $y$ increases.' However, we can't say 'at a constant rate' as the line is a curve and not a straight line.

In your conclusion you should also say whether any prediction made before you started your investigation was supported by the data collected and displayed. You should use the science that you have learned to try to explain **why** any pattern you spot happened. The more scientific detail you can include the better the mark you can get.

# Evaluating your investigation

At the end of your investigation you should reflect on how well your investigation went. Could you have improved the method to get more accurate data? Here are some questions you should ask yourself:

- Were your data reproducible? Did other people doing the same investigation get the same results?
- Were your data repeatable? Did you decide to repeat readings? If so were they precise (grouped closely together)?
- Was your investigation as fair as possible? How did you take into account any variables that were difficult (or impossible) to control? Was sample size an issue?
- What types of errors might have been made in your investigation? How could you have improved your method if you had the chance? Why would your suggestions have produced better quality data?
- Did you have enough data to draw conclusions? Comment on the range and interval of the variables in your investigation. Did they limit the conclusion you made? If this was an issue, how could you improve this in future?

## Activity

### Practising evaluation

Look at the pictures of students doing an investigation. They wanted to find out which Bunsen flame was hottest.

The starting temperature is 18 °C.

I've got some fresh water. Now let's heat the beaker for a minute, each time with the air-hole open by different amounts.

OK, the temperature with this flame is now 19 °C.

From what you can see in the pictures, comment on any ways they could have improved their investigation.

## Summary questions

1 What do we mean by:
   a a positive linear relationship
   b a negative linear relationship
   c a directly proportional relationship?

2 Design a checklist of statements to use as a reminder of what to do when evaluating your next investigation.

AQA Examiner's tip

Throughout your Controlled Assessment unit it is important that you use and understand some important keywords:

**Accuracy** – an accurate measurement is one that is close to the true value.

**Fair test** – a test where only the independent variable has been allowed to affect the dependent variable. All other variables are kept constant.

**Precision** – this is related to the degree of scatter from the mean. The more tightly grouped a set of repeat readings is, the more precise it is.

**Repeatable** – results that when repeated, give a consistent set of measurements.

**Reproducible** – when checking your data with that collected by someone else doing the same investigation, the results are consistent.

## Key points

- A conclusion should say if there is any pattern in the data collected.

- Use the science you have learned to try to explain why any pattern you spot happened.

- A conclusion should also say whether any prediction made before you started your investigation was supported by the data collected.

- At the end of your investigation you should reflect on how well your investigation went.

# Glossary

## A

**Accuracy** This tells us how near the true value a measurement is.

**Adaptations** Special features that make an organism particularly well suited to the environment where it lives.

**Addicted** The body's dependency on a chemical (drug).

**Alkane** Saturated hydrocarbon with the general formula $C_nH_{2n+2}$, for example, methane, ethane and propane.

**Allele** A version of a particular gene.

**Allergy** The immune system's response to a usually harmless substance.

**Alpha particle** Two protons and two neutrons emitted from the nucleus.

**Alveoli** The air sacs found in the lung.

**Anomalous results** Results that do not match the pattern seen in the other data collected or are well outside the range of other repeat readings. They should be re-tested and if necessary discarded.

**Antibiotics** Drugs that destroy bacteria inside the body without damaging human cells.

**Antibodies** The chemicals which 'deactivate' microorganisms, stopping them from causing disease.

**Anti-toxins** The chemicals which destroy the poisonous toxins that some microorganisms make.

**Atmosphere** The relatively thin layer of gases that surround planet Earth.

**Atomic number** The number of protons (which equals the number of electrons) in an atom. It is sometimes called the proton number.

**Atom** The smallest part of an element that can still be recognised as that element.

**Auxin** A plant hormone that controls the responses of plants to light (phototropism) and to gravity (gravitropism).

## B

**Bar chart** A chart with rectangular bars with lengths proportional to the values that they represent. The bars should be of equal width and are usually plotted horizontally or vertically. Also called a bar graph.

**Beta particles** High speed electrons emitted from the nucleus.

**Big Bang** The theory that the Universe was created in a massive explosion (the Big Bang) and that the Universe has been expanding ever since.

**Biodegradable** Materials that can be broken down by microorganisms.

**Biofuel** Fuel made from animal or plant products.

**Biomass** Plant and animal material used as a fuel or the mass of all living material found in an ecosystem.

**Brittle** Not flexible, will break if it is bent.

## C

**Calcium carbonate** The main compound found in limestone. It is a white solid whose formula is $CaCO_3$.

**Carbohydrates** A food group which provides a rich source of energy.

**Carbon cycle** The cycling of carbon through the living and non-living world.

**Carbon monoxide** A toxic gas whose formula is CO.

**Carriers** Individuals who are heterozygous for a faulty allele that causes a genetic disease in the homozygous form.

**Cement** A building material made by heating limestone and clay.

**Central nervous system (CNS)** The central nervous system is made up of the brain and spinal cord where information is processed.

**Ceramics** Hard solid materials with a high melting point, such as china or brick.

**Classification** The sorting of organisms or objects into groups based on similar features.

**Clones** Offspring produced by asexual reproduction which are identical to their parent organism.

**Combustion** The process of burning.

**Community** All the living organisms present in a habitat.

**Composites** Combinations of two or more materials.

**Compound** A substance made when two or more elements are chemically bonded together. For example, water ($H_2O$) is a compound made from hydrogen and oxygen.

**Compression** Squeezing together.

**Concrete** A building material made by mixing cement, sand and aggregate (crushed rock) with water.

**Conduction** Transfer of energy from particle to particle in matter.

**Convection** Transfer of energy by the bulk movement of a heated fluid.

**Convection currents** The circular motion of matter caused by heating in fluids.

**Core** The centre of the Earth.

**Cosmic microwave background radiation** Electromagnetic radiation that has been travelling through space ever since it was created shortly after the Big Bang.

**Crude oil** A fossil fuel extracted from deep underground.

**Crust** The outer solid layer of the Earth.

## D

**Decomposers** Microorganisms that break down waste products and dead bodies.

**Diabetes** A condition where sufferers cannot control their blood glucose levels.

**Displacement reaction** A chemical reaction in which one chemical replaces another.

**DNA** Deoxyribonucleic acid, the material of inheritance.

**Dominant allele** The characteristic that will show up in the offspring even if only one of the alleles is inherited.

**Doppler effect** The change of wavelength (and frequency) of the waves from a moving source due to the motion of the source towards or away from the observer.

**Droplet infection** A way of spreading infectious diseases through the tiny droplets full of pathogens, which are expelled from your body when you cough, sneeze or talk.

**Drug** A chemical which causes changes in the body. Medical drugs cure disease or relieve symptoms. Recreational drugs alter the state of your mind and/or body.

# E

**Ecosystem** All the living and non-living matter present in an area.

**Electrical energy** Energy transferred by the movement of electrical charge.

**Electrode** An electrical conductor used in electrolysis.

**Electrolysis** The breakdown of a substance containing ions by electricity.

**Electrolyte** A liquid, containing free-moving ions, that is broken down by electricity in the process of electrolysis.

**Electromagnet** A magnet that turns on when a current flows.

**Electromagnetic spectrum** A set of radiations that have different wavelengths and frequencies but all travel at the same speed in a vacuum.

**Electromagnetic wave** Electric and magnetic disturbances that transfer energy from one place to another. The spectrum of electromagnetic waves, in order of increasing wavelength, is as follows: gamma and X-rays, ultraviolet radiation, visible light, infrared radiation, microwaves, radio waves.

**Electron** A tiny particle with a negative charge. Electrons orbit the nucleus in atoms or ions.

**Electroplating** The process of depositing a thin layer of metal on an object during electrolysis.

**Element** A substance made up of only one type of atom. An element cannot be broken down chemically into any simpler substance.

**Embryo** A developing baby from fertilisation until eight weeks.

**Enzyme** Protein molecule that acts as a biological catalyst.

**Ethanol** Chemical found in alcoholic drinks and biofuels such as gasohol, its chemical formula: $C_2H_5OH$.

**Ethene** An alkene with the formula $C_2H_4$.

**Eutrophication** The process of excess nutrients entering a water system, resulting in excessive algal growth.

**Evidence** Data which have been shown to be valid.

**Evolution** The process of slow change in living organisms over long periods of time as those best adapted to survive breed successfully.

**Evolutionary tree** Model of the evolutionary relationships between different organisms based on their appearance, and increasingly, on DNA evidence.

**Extinct** The permanent loss of all the members of a species.

# F

**Fair test** A fair test is one in which only the independent variable has been allowed to affect the dependent variable.

**Fertilisers** Chemicals which add minerals to the soil.

**Fetus** A developing baby from the eighth week of conception.

**Fluids** A liquid or a gas.

**Food webs** Diagrams showing how food chains are interlinked.

**Fossil fuel** Fuel obtained from long-dead biological material, coal, oil or gas.

**Fractional distillation** A way to separate liquids from a mixture of liquids by boiling off the substances at different temperatures, then condensing and collecting the liquids.

**Free electrons** Electrons that move about freely inside a metal and are not held inside an atom.

**Frequency** The number of wave crests passing a fixed point every second.

# G

**Galaxy** A group of billions of stars.

**Gamma radiation** Electromagnetic radiation emitted from unstable nuclei in radioactive substances.

**Gamma rays** High energy electromagnetic radiation.

**Gene therapy** The insertion of a desired gene into a person's genetic material to treat/cure a disease.

**Generator** A machine that produces voltage.

**Genetic engineering/modification** A technique for changing the genetic information of a cell.

**Geothermal energy** Energy from hot underground rocks.

**Glands** Organs that produce fluids such as hormones or sweat.

**Global warming** The increasing of the average temperature of the Earth.

**Gravitropism** Response of a plant to the force of gravity controlled by auxin.

**Greenhouse effect** A planet's own atmosphere insulates and warms the planet.

**Greenhouse gas** Gases, such as carbon dioxide and methane, which absorb infrared radiated from the Earth, and result in warming up the atmosphere.

# H

**Habitat** The place where an organism lives.

**Haemoglobin** The component of red blood cells which binds to the oxygen molecule.

**Hertz** The unit of frequency.

**Homeostasis** The maintenance of constant internal body conditions.

**Hydrocarbon** A compound containing only hydrogen and carbon.

**Hydroelectric** Using falling water to generate electricity.

# I

**Immunisation** Giving a vaccine that allows immunity to develop without exposure to the disease itself.

**Immunity** Antibodies that present in the body 'fight off' a microorganism before they can cause disease.

**Incomplete combustion** When a fuel burns in insufficient oxygen, producing carbon monoxide as a toxic product.

**Indicator species** Lichens or insects that are particularly sensitive to pollution and so can be used to indicate changes in the environmental pollution levels.

**Infrared radiation** Electromagnetic waves between visible light and microwaves in the electromagnetic spectrum.

**Inherited** Passed on from parents to their offspring through genes.

**Insulin** The hormone which controls blood sugar levels.

**Intensive farming** Farming that produces as much food as possible in the space available.

**Ion** A charged particle produced by the loss or gain of electrons.

**Ionising radiation** Radiation that can knock electrons (into or) from atoms, changing them into ions.

# J

**Joule (J)** The unit of energy.

# K

**Kinetic energy** Energy of a moving object due to its motion; kinetic energy (in joules, J) = mass (in kilograms, kg) $\times$ (speed)$^2$ (in m$^2$/s$^2$).

# L

**Limestone** A sedimentary rock with many uses in building and manufacture.

**Line graph** Used when both variables are continuous. The line should normally be a line of best fit, and may be straight or a smooth curve. (Exceptionally, in some (mainly biological) investigations, the line may be a 'point-to-point' line.)

**Longitudinal wave** Wave in which the vibrations are parallel to the direction of energy transfer.

# M

**Mantle** The layer of the earth between its crust and its core.

**Mass number** The number of protons plus neutrons in the nucleus of an atom.

**Metal ores** Rocks containing enough metal compounds to be worth extracting.

**Microbes** Microscopic organisms – also known as microorganisms.

**Microorganisms** Bacteria, viruses and other organisms that can only be seen using a microscope.

**Microwave** Part of the electromagnetic spectrum.

**Mixture** When some elements or compounds are mixed together and intermingle but do not react together (i.e. no new substance is made). A mixture is *not* a pure substance.

**Molecule** A group of atoms bonded together, foe example, $PCl_5$.

**Monomers** Small reactive molecules that react together in repeating sequences to form a very large molecule (a polymer).

**Moon** A planet's natural satellite.

**Mortar** A building material used to bind bricks together. It is made by mixing cement and sand with water.

**Mould** A type of fungus.

**MRI scanner** Magnetic resonance imaging scanner.

**Mutate** To change the genetic material.

**Mutation** A change in the genetic material of an organism.

# N

**National Grid** The network of cables and transformers used to transfer electricity from power stations to consumers (i.e. homes, shops, offices, factories, etc.).

**Natural selection** The process by which evolution takes place. Organisms produce more offspring than the environment can support so only those which are most suited to their environment – the 'fittest' – will survive to breed and pass on their useful characteristics.

**Negative feedback** A system by which any changes which affect the body are reversed, and returned to normal.

**Neuron(s)** Basic cells of the nervous system which carry minute electrical impulses around the body.

**Neutron** A dense particle found in the nucleus of an atom. It is electrically neutral, carrying no charge.

**Nicotine** The additive chemical present in tobacco.

**Non-renewable** Something that cannot be replaced once it is used up.

**Nuclear fission** The process in which certain nuclei (uranium-235 and plutonium-239) split into two fragments, releasing energy and two or three neutrons as a result.

**Nucleus** The very small and dense central part of an atom which contains protons and neutrons.

# O

**Ore** Rock which contains enough metal to make it economically worth while to extract the metal.

**Oxidised** A reaction where oxygen is added to a substance (or when electrons are lost from a substance).

**Oxo-degradable** A material that breaks down when exposed to air.

# P

**Pathogen** Microorganism which causes disease.

**Payback time** The time taken for something to produce savings to match how much it cost.

**Periodic table** An arrangement of elements in the order of their atomic numbers, forming groups and periods.

**Photo-degradable** A material that breaks down when exposed to light.

**Photosynthesis** The process by which plants make food using carbon dioxide, water and light energy.

**Phototropism** The response of a plant to light, controlled by auxin.

**Phytomining** The process of extraction of metals from ores using plants.

**Planet** A large object that moves in an orbit round a star. A planet reflects light from the star and does not produce its own light.

**Plasmids** Extra circles of DNA found in bacterial cytoplasm.

**Platelets** Fragments of cells used in blood clotting.

**Polymer** A substance made from very large molecules made up of many repeating units, for example, poly(ethene).

**Polymerisation** The reaction of monomers to make a polymer.

**Population** The members of one species living in a habitat.

**Potential difference** A measure of the work done or energy transferred to e.g. a lamp by each coulomb of charge that passes through it. The unit of potential difference is the volt (V).

**Power** The energy transformed or transferred per second. The unit of power is the watt (W).

**Precision** The close grouping of repeat readings.

**Predator** An animal which preys on other animals for food.

**Prey** An animal that a predator kills for food.

**Producers** Organisms which make their own food by the process of photosynthesis.

**Product** A substance made as a result of a chemical reaction.

**Proton** A tiny positive particle found inside the nucleus of an atom.

## R

**Radioactivity** Unstable nuclei which emit ionising radiation that changes into more stable nuclei.

**Radon** A naturally occurring radioactive gas.

**Rarefaction** Stretching apart.

**Reactant** A substance we start with before a chemical reaction takes place.

**Reactivity series** A list of elements in order of their reactivity. The most reactive element is put at the top of the list.

**Recessive** The characteristic that will show up in the offspring only if both of the alleles are inherited.

**Red-shift** Increase in the wavelength of electromagnetic waves emitted by a star or galaxy due to its motion away from us. The faster the speed of the star or galaxy, the greater the red-shift is.

**Reducing agent** A chemical used to remove the oxygen from a metal oxide.

**Renewable energy** Energy from sources that never run out including wind energy, wave energy, tidal energy, hydroelectricity, solar energy and geothermal energy.

**Repeatable** A measurement is repeatable if the original experimenter repeats the investigation using the same method and equipment and obtains the same results.

**Reproducible** A measurement is reproducible if the investigation is repeated by another person, or by using different equipment or techniques, and the same results are obtained.

**Resistance** Resistance (in ohms, $\Omega$) = potential difference (in volts, V) ÷ current (in amperes, A).

**Respiration** The process by which food molecules are broken down to release energy for the cells.

## S

**Selective breeding** The process of selecting desired characteristics amongst organisms, through controlled breeding.

**Sense organs** Collection of special cells known as receptors which respond to changes in the surroundings (e.g. eye, ear).

**Solar system** The Sun, planets and all the other objects that orbit the Sun.

**Spectroscopy** The analysis of starlight using a spectroscope.

**Star** A massive ball of hot gases in space that radiates energy.

**Stem cells** Cells that have the ability to grow into any type of cell in the body.

**Step-down transformers** Used to step the voltage down, for example, from the grid voltage to the mains voltage used in homes and offices.

**Step-up transformers** Used to step the voltage up, for example, from a power station to the grid voltage.

**Superconductivity** The property of having nearly zero resistance.

**Superconductor** A material that has nearly zero resistance.

**Sustainability** Balancing the needs of industry against care for the environment.

## T

**Tectonic plates** The huge slabs of rock that make up the Earth's crust and top part of its mantle.

**Thermal decomposition** The breakdown of a compound by heat.

**Total internal reflection** This is when light repeatedly reflects off the inside surface (for example, of a glass fibre).

**Trophic level** The position in a food chain.

**Tropism** The growth in response to an environmental stimulus.

**Turbine** A machine that uses steam or hot gas to turn a shaft.

## U

**Ultrasound** Sound with a frequency above 20 000 Hz.

**U-value** A measure of the heat flow through a material.

## V

**Vaccine** The dead or inactive pathogen material used in vaccination.

**Variable – dependent** The variable for which the value is measured for each and every change in the independent variable.

**Variable – independent** The variable for which values are changed or selected by the investigator.

**Variation** Differences within a species.

## W

**Watt (W)** The unit of power.

**Withdrawal symptoms** The symptoms experienced by a drug addict when they do not get the drug to which they are addicted.

## X

**X-ray** A form of electromagnetic radiation.

# Index

Published in 2011 by:
Nelson Thornes Ltd
Delta Place
27 Bath Road
CHELTENHAM
GL53 7TH
United Kingdom

11 12 13 14 15 / 10 9 8 7 6 5 4 3 2 1

A catalogue record for this book is available from the British Library

ISBN 978 1 4085 0835 0

Cover photograph: Getty Images/Tanya Constantine

Illustrations by GreenGate Publishing, with additional artwork by Wearset

Index created by Indexing Specialists

Page make-up by GreenGate Publishing

Printed and bound in Spain by GraphyCems

## Acknowledgments

**Alamy:** /Ace Stock Limited 10.6.4, /A J and H Evans 15.4.3, / AW 143b, /David J Green - work themes 15.8.4, /David Young-Wolff 9.4.1, /Galaxy Picture Library 1.1.1, /Global Warming Images 14.4.3, /Juice Images p143t, /Photofusion Picture Library 10.4.2, /Vario Images GmbH & Co.KG 2.5.4; **Corbis**: / DK Limited 12.2.3; **FLPA:** /Willem Kolvoort/FN/Minden 14.3.3; **Fotolia:** /Avesun 7.5.4, /Codreanu Mihai p3bl, /Elridge 9.3.1, /gwimages 9.7.2, /hrv p75br, /Igor Kaliuzhnyi 10.4.1, /Kim Warden p3tr, /kmit p72l, /Laurentiu Iordache 2.5.3, /Marc Tielemans 7.3.4r, /Mariusz Blach 12.3.1, /Nialat 3.2.2r, /Serbor 4.6.1, /Shariff Che'Lah 2.7.2, /Sinisa Botas 2.6.5, /Sommersby 15.4.4, /sumos p3tl, /Yang yu 4.6.2, /YellowCrest 7.5.6; **Getty Images:** /John Lawrence 9.1.3, /Matthew Piper 7.3.4l, /Mike D Kock 7.7.2, /Yann Layma 2.7.1r; **iStockphoto:** 1.1.2, 2.1.2 (all), 2.1.3l, 2.1.4 (all), 2.3.4, 2.4.1, 2.4.2, 2.4.5, 2.4.6, 2.5.1, 2.6.3, 2.6.4, 2.7.1l, 2.8.6, 3.1.1, 3.2.2l, 3.2.4, 3.2.5, 3.2.6, 3.2.7, 3.2.8, 3.2.9, 3.3.1, 3.4.1, 3.4.3, 3.4.4, 4.1.1, 4.1.3, 4.3.4, 4.4.1, 4.5.2 (all), 4.6.3, 5.1.2, 5.2.3, 5.4.3, 6.2.2, 7.1.2, 7.1.3, 7.3.1, 7.3.2, 7.4.1, 7.4.2, 7.5.5, 7.5.7 (all), 7.6.1, 7.6.2, 7.6.4, 7.6.5, 7.6.7, 7.6.8, 8.1.1, 8.7.2, 8.7.3, 9.1.1, 9.3.4, 9.5.1, 9.6.2, 9.6.4, 9.7.1, 10.2.1, 10.6.5, 11.4.1, 11.4.3, 11.6.3, 12.3.2, 12.6.3, 13.2.1, 13.2.3, 13.4.1, 13.4.4, 14.1.1, 15.2.2, 15.2.3, 15.3.1, 15.3.3, 15.3.4, 15.5.2 (all), 15.5.3, 15.5.4, 15.6.1, 15.7.12, 15.9.1, p42, p43t, p43b, p57l, r, p74, p107tl, tr, bl, p140b, p140t, p197tr, p162, p163t, p163m, p223bl; **James Hayward:** 7.3.5; **Jim Margeson:** 7.3.3; **Martyn Chillmaid:** 2.3.5, 5.8.2, 5.8.3, 5.8.4, 6.2.4, 10.1.2, 12.4.4, 12.4.5, 15.7.10; **Natural Building Technologies (NBT):** 7.7.5; **Nissan Motor Manufacturing (UK) Ltd:** 12.4.3, 12.6.2; **PA Photos:** p223tr; **Photodisc 31 (NT):** p223br; **Photolibrary.com:** 6.2.3, 7.3.4m, 15.3.5; **Rex Features:** 3.1.2, p197tl, /John Freeman 6.5.3; /Solent News p107br; **Science Photo Library:** 1.4.1, 3.5.1, 10.5.2, 11.6.2, /Adam Hart-Davis 2.2.1, /Adrienne Hart-Davis 7.5.3, /Alan Sirulnikoff 7.7.4, /Alex Bartel 7.6.6, 12.6.1, /Andrew Lambert Photography 8.1.4, / Astrid & Hanns-Frieder Michler 10.3.1, /Biomedical Imaging Unit, Southampton General Hospital 10.2.3, /Brian Gadsby 8.8.3, /BSIP, Sercomi 13.3.1, /Carlos Dominguez 15.7.9, / Charles D Winters 2.8.3, 12.1.1, /Charlotte Raymond 10.2.2, / CNRI 10.6.3 (both), /Cordelia Molloy 7.6.3, /Cristina Pedrazzini 5.4.2, /Custom Medical Stock Photo 10.5.1, 11.8.1, /Daniel Sambraus,Thomas Luddington 2.1.3r, /David McCarthy 3.5.3, p43m, /David Parker 3.4.2, /Dr Arthur Tucker 15.5.1, /Dr Gopal Murti 6.1.1, /Dr John Brackenbury 11.2.2, /Dr Kari Lounatmaa 11.1.2, /Dr P Marazzi 5.6.2, 5.6.3, 11.8.2, 12.3.3, /Duncan Shaw 14.3.2, /Eckhard Slawik 1.3.1, /E R Degginger p223tl, / European Southern Observatory 1.1.3, /Eye of Science 15.6.5, /Geoff Tompkinson p197br, /George Steinmetz 5.5.4, /Health Protection Agency 15.9.4, /Ian Hooton p75ml, /James King-Holmes 15.8.5, /Jim Varney 11.2.1, /Makoto Iwafuji/Eurelios p197bl, /Mark Clarke 5.8.1, /Martin Bond 8.6.2, 8.7.4, 9.2.1, 12.5.2, p196t, /Martyn F Chillmaid 9.2.1, 9.2.2, 9.3.2, 9.4.3, 11.7.5, 12.3.4, 15.7.8, 15.7.11, p142, /Massimo Brega/Eurelios p196b, /Mauro Fermariello 9.7.3, /Maximilian Stock Ltd 13.3.2, /Michael Donne p3br, /Pascal Goetgheluck 12.4.1, /Pasieka 11.1.1, p75tr, /Patrice Latron/Look at Sciences p163b, /Ph. Plailly / Eurelios 13.4.3, /Phillip Wallick / Agstockusa 14.1.3, / Professor Miodrag Stojkovic 6.5.2, /Ria Novosti 8.5.2, /Robert Brook 2.10.1, 8.8.2, 14.4.2, /Sam Ogden 8.1.2, /Science Source 1.4.2, /Sheila Terry 7.5.2, /Simon Fraser 11.4.2, p222, /Sinclair Stammers 11.7.1, /SIU 13.4.5, /Sovereign, ISM 12.4.2, /Spencer Grant 12.5.3, /Steve Allen p140bm, /Steve Gschmeissner 6.1.2, 15.8.2, /Sue Ford 14.1.2, /TEK Image 9.3.3, 15.7.1, /Thedore Gray, Visuals Unlimited 2.6.2, /Tony McConnell 15.4.1, /U.S. Dept. of Energy 8.6.3, 12.5.1, /USCG 8.4.1, /Will & Deni McIntyre 6.4.1, /ZEPHYR 12.5.4; **www. glofish.com:** 13.2.2; **www.pelamiswave.com:** 8.7.5, p140tm.

Figures 8.5.1, 14.4.1 and 15.9.2: Material adapted from The Department for Environment, Food and Rural Affairs under the Open Government Licence www.nationalarchives.gov.uk/doc/open-government-licence/